DECARBONISE DIGITAL

FACTS. METHODS. ACTION.

BY ERIC ZIE

'Decarbonise Digital : Facts. Methods. Action.'
By Eric Zie

ISBN 978-1-3999-6476-0

Published on Amazon KDP
I asked, there is no recycled paper option available from Amazon KDP
First Edition, 2023

www.decarbonise.digital

With thanks to Rachael Walker and Bobby Keetch for proofreading and your corrections and suggestions. The book reads better for your time and patience.

About the Author

Eric holds a Masters from the London School of Economics and is a career technologist with over twenty-five years spent in Europe and Asia building, designing, and running digital services. In recognition of his efforts to decarbonise digital, he has been named Computing's Digital Ambassador of the Year for 2023. He is World Green Technology Leader and the recipient of the Google Brain Award for the use of artificial intelligence.

He was born in Hong Kong and moved to the United Kingdom in June 1976. During his first summer in London, his earliest memory is of arriving in vast, open green spaces and observing swarms of flying ants. Since then, he has discovered that flying ants enjoy long periods of intense heat, and he often explains this by showing the NASA heat map[1] of the world from that summer, which highlights hot spots in western Europe (see left image below). He also compares this map to the June 2022 equivalent. You can clearly see the difference in temperatures, where the global anomaly has increased from -0.15°C to +0.93°C. He is driven to take action because the worst effects of climate change have accelerated so significantly in his own lifetime.

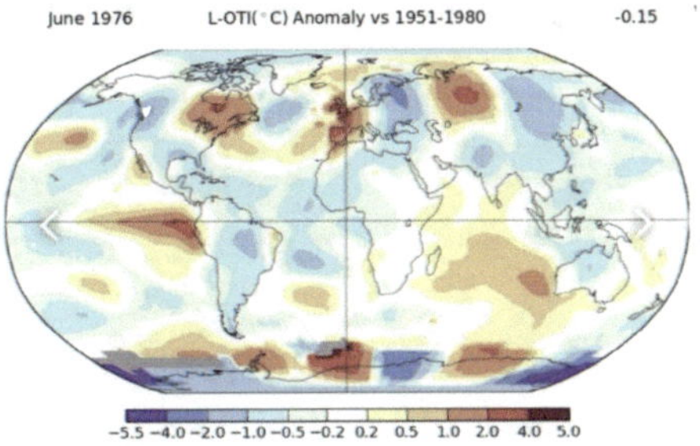

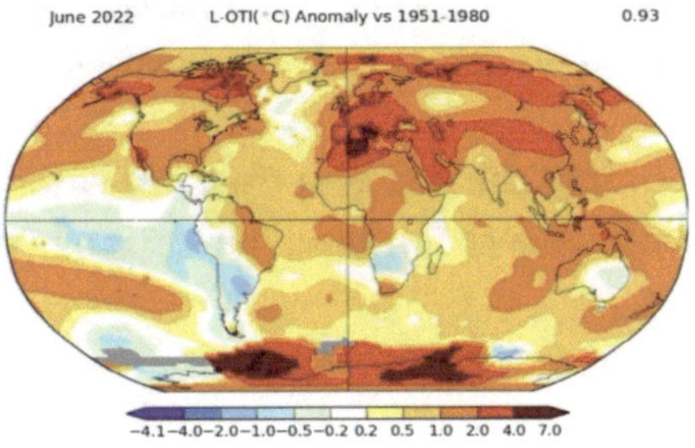

---

[1] Source: Nasa Earth Observations, neo.gsfc.gov

Eric is the CEO and Founder of GoCodeGreen, a UK-based, independent, and global leader in ClimateTech with a mission to help decarbonise the digital world in which we all live. As part of a coalition taking digital action, GoCodeGreen is partnered with the United Nations agency for digital technology, the International Telecommunication Union (ITU), standards bodies, and the World Bank.

GoCodeGreen is nominated for the Earthshot Prize in 2023, and is a certified B Corporation committed to the triple bottom line of balancing people, planet, and profit.

Eric and his family live in Suffolk with Belle, a Sprocker spaniel who is Eric's best (non-human) friend. He's an avid tennis player, triathlete, sourdough baker, open-fire cook and martini drinker (gin, stirred not shaken).

A note to the reader

Writing this book wasn't part of the plan; I'm just a technologist trying to do some good, and the idea of putting pen to paper only came about when I was asked to recommend reading material on decarbonising digital and sustainable IT for a group of university students. I was stuck, having found what was available too focused on simply observing the problem or lacking sufficient detail to be useful. This is my best effort to share with others what I have learned over the past few years as a practitioner and someone wanting to tackle the problem. We have so little time to fight climate change, so if you are able to use any of the information and insight in this book to make a positive difference, I will be delighted, and you will have made the book and all the work that led to it worthwhile.

I simply ask that you treat my writing in the spirit in which it was undertaken, with pace rather than perfection in mind.

Thank you for taking action.

Eric
November 2023

For my brother, Jeff
(1969-2021)
You helped me more than you know.
I'll keep trying to do some good.

Two years ago, on September 22, 2021, my brother, Jeff, decided to end his own life. I still don't know why. I've been angry, disappointed, and so deeply sad that he didn't reach out to me through what must have been the most difficult time in his life.

Two years later, I try to remember him by his smile and his laugh and by how serious he was about trying to do good to help us be better in society and for the planet. He was a campaigner and a supporter of good causes, including mental health awareness.

I am left with the idea that he decided his time was done and that, for his own reasons, he no longer had the energy to fix the wrongs he saw around him. Every day since September 22nd, 2021, I wish we had found that time to talk about how he was feeling and, maybe together, we might have had the energy to fight the fights that needed to be fought. We did speak about climate change and the work he was doing ahead of COP26, and I remain inspired to play my part, now with my own mission but sadly without him beside me.

I never got the chance to listen and talk about his pain; I didn't make the opportunity to help, and that remains the biggest regret I have in life.
I miss you, brother.

Everybody, please find time to be kind and to talk. A percentage of profits from the sale of this book will be donated to Mind UK to help fund research and support their campaigns for infants, children, young people, and their families affected by mental illness.

Foreword

Since the nineteenth century and industrialisation, carbon emissions have been growing at an alarming rate. Today, after the global pandemic and a lot of focus on the environment, we are just waking up to the seriousness of the situation. At this moment, I was lucky to meet Eric Zie, a leading light in the field, and since our meeting, we have been working together to see how we can make a positive difference.

Eric has enthusiasm and energy, along with clear plans and approaches for reducing carbon and moving us towards net zero in a meaningful and straightforward way. This does mean looking at everything we are doing, but also looking at what we need to do as the world moves from pandemic mode to understanding how AI will transform our world and lives.

While there is growing interest in his topic, there is still a great deal of noise and uncertainty regarding the veracity of data and how to utilise it. This book is based on the author's experiences and knowledge gained from using technology. In it, he envisions laying the foundation for more sustainable digital growth in the future while working towards decarbonising our processes and systems.

The purpose of any foreword is to persuade the reader that the book they are holding is worth their time to read and that the author is worthy of their attention. You are utilising the expertise of a distinguished digital ambassador and global leader in his profession. He is a positive disruptor, and if there was ever a time when we needed his agitation to force change, it is now. If you are concerned about climate change and have any connection to the digital world, then you should read this book and apply it to your workplace.

Sara de Freitas
Author, BT Professor and Director of the Digital Futures Institute
October 2023

Index

Preface

Look around, and it is clear that we now live in an increasingly digital world. Everywhere we turn, we hear talk of Future Cities, Smart Technology, Artificial Intelligence, Big Data, and Generative AI. Technology and digital solutions are here and growing, with the social, educational, health, and business value of technology unquestionable. But behind the scenes, this growth has a carbon cost that has remained invisible to most of us.

The Information and Communication Technology (ICT) sector is now estimated to account for up to 5 percent of global greenhouse gas emissions[2] and is predicted to consume over 20 percent of global energy by 2030[3]. A much-quoted comparison that continues to shock is that data centres running the software we so depend on now have the same carbon impact as the global aviation industry.

Everything, everywhere, all at once.

The recent IPCC (the Intergovernmental Panel on Climate Change, an organisation established by the United Nations to assess the science related to climate change) synthesis report, the conclusion of its sixth assessment cycle, gave us an updated call to action. The world's leading scientists have once again shown us that we have no time to spare and that, as António Guterres, the Secretary General of the United Nations, stated, we now must do "everything, everywhere, all at once" if we are to hold to the 1.5°C temperature change set by the Paris agreement[4]. By the time the next IPCC assessment cycle completes, it may well be 2030. It is the same timeframe we

---

2 'The EU climate strategy for the ICT sector', European Parliament, Sept. 2022

3 Enerdata, 'Executive Briefs', 2018

4 Please go to the 'Stop Press' update on the very last page of this book.

have to make a difference to avoid reaching the tipping point, the critical threshold that, if crossed, will lead to large and often irreversible changes in the climate system.

We are now able to hold a mirror up, and what we all now need is for you to look at yourselves and embrace the opportunity to make an even better, faster, more efficient, planet-friendly version of your amazing digital solutions.

Digital and technology underpin almost everything we do; they are everywhere, and the time is now to start focusing on reducing their energy consumption. Having spent the last two years engaging with some of the world's leading cloud service providers, software companies, digital leaders, technology consultancies, and system integrators, it is now my strong opinion that we have an opportunity ahead of us, one that could see us take the technology sector to the forefront of decarbonisation.

In our digital world, we must move forward, our actions must be progressive, and we must be able to utilise the advancements in technology to create the efficiencies needed to support the growth of digital.

The technology sector is full of some of the smartest people around. We are doers. Our ingenuity, innovation mindset, and sheer determination to solve problems make us a highly capable group. We must stop thinking that taking action means we will need to, in some way, slow things down and that we have to regress to some version of the past that might have been less advanced and polluting. In our digital world, we must move forward, our actions must be progressive, and we must be able to utilise the advancements in technology to create the efficiencies needed to support the growth of digital. But we should recognise that we can only meet the needs of societies and businesses for more technology if we base this growth on a more sustainable IT sector.

Look out for this sign

When you see this symbol throughout the book it indicates an opportunity to complete a practical exercise or use a resource that may be helpful in your digital decarbonisation journey. Please take the time to use these resources as they help to bring the knowledge to life.

Introduction

When I started thinking seriously about writing this book, I knew I needed a structure to organise my thoughts. My mind works in pictures, so I imagined it would go something like this[5]:

[5] Artist: Jonath Prac.

So a manga storyline became my guide, and I tell you now that I will jump from macro to micro level and from one large topic to another through the book. This is a reflection of my personal learning journey over the past two years, and after careful consideration, I've decided to stick and go with it as a structure. Let me now walk you through how everything came to be and why I believe that building the knowledge in these steps is useful.

Structuring the book

After spending eighteen months researching, experimenting, designing, and building the calculation platform at GoCodeGreen, I undertook an extensive six-month beta phase where the initial product was used in anger with clients in what is best described as a true front-line test-and-learn approach. It became clear early on that I had to consider the non-technical enablers needed to make GoCodeGreen successful and to listen to the feedback and challenges from a variety of clients, all of whom helped shape each subsequent release of the platform.

Education was identified early on as one of those crucial enablers. I would typically spend half of my time explaining to audiences why they should care about the carbon and climate impacts of digital before even starting to talk about my solution. No surprise, given that the majority of people at that time were still more concerned with switching to LED light bulbs in their offices as their primary means of lowering carbon emissions. The impact of digital wasn't even on their radar. I dedicated a large amount of time to bringing this impact to life through examples, such as one of my favourites, High School Musical, and Olivia Rodrigo[6], and these became central to the

---

[6] It goes like this: "You look like a High School Musical fan? Well, if you were, then you'd know that Olivia Rodrigo launched her song, Drivers Licence, and it went viral. The number of streams on Spotify in a year used the energy, resulting in emissions of over 3,000 metric tonnes of CO2. You'd need nearly 200,000 trees to sequester that amount of carbon." Cue gasps from the audience.

majority of the pitches I gave. It was evident that there was a lack of understanding and awareness of the problem, but there was also a strong desire to learn more.

This is when I realised that everyone would benefit from gaining enough foundational knowledge to enable them to appreciate the challenge and opportunity associated with decarbonising digital. As a result, I decided to create my first learning course, and the research undertaken and information gained are the basis for **Part One, where I will explain key climate terminology, establish the context for digital impact, introduce the importance of Information and Communication Technology (ICT), and cover what I now refer to as my 'recipe for change'.**

As the GoCodeGreen measurement products evolved, I also wanted to assess the readiness of the technical teams responsible for developing the software products we were evaluating. Ultimately, it would be these people who would need to make the changes needed to enhance the efficiency of the software and platforms. It was not surprising that the self-assessments I created for engineers consistently revealed 'poor' or 'fair' knowledge of green coding principles and practices. What was surprising was how quickly the majority of clients would ask, "What can we do about it? Where can we go to help my team get the skills and knowledge to make a difference?". **The content in Part Two became the basis of a series of much more in-depth and technically focused learning courses that explore the principles and techniques of 'sustainable software engineering' and how to implement them in real-world scenarios.**

As you might expect, many of the digital products we ended up evaluating were retrospective in nature, i.e., the software had been developed and was already leaving a carbon footprint and having an impact on the environment. That's when I developed what we now refer to as our 'design stage simulator', with the intention of assisting teams prior to coding by simulating the carbon impact of their new digital product based on decisions made during the design discovery and problem solving phases. The idea of being more proactive and forward-thinking became a real focal point for me. The simplicity of not 'adding to the problem' by making smarter design decisions that would result in greener digital products made perfect sense and would be so much more effective than attempting to remediate an energy-inefficient product after it had already been released and installed. **Part Three was used to create the 'Introduction to Green Design' learning course I developed to cover how design-based thinking and 'green design' principles must be at the core of all new product design, as well as the practical techniques that should be applied to all new digital products.**

As the platform's usage continued to increase, two prominent themes began to emerge. First, upon seeing the results of individual digital product assessments, which were great for application teams to target action at the product level, IT leaders and decision makers wanted to jump straight to aggregating the findings across multiple assessments in order to identify larger levers that they could pull in order to achieve scaled carbon reductions. For them to begin constructing a sustainable IT strategy for their organisations, they also wanted a deeper understanding of the broader context of their IT's impact beyond software alone. The second theme was that customers frequently preferred to start their decarbonisation journeys with our learning and self-assessment tools before measuring their digital products and services. The concept of preparing their workforce to take action was typically easier to sell within their organisations and could serve as a catalyst for change and to gather support for more investment.

These two themes resulted in a further and significant round of product development at GoCodeGreen, leading me to build a series of screening tools to span the three main categories of Information and Communication Technology (ICT), i.e., desktop managed services, network and telecoms, and data centre and cloud services, as well as develop a broader range of self-assessments to include designers and IT leaders. The learning and work undertaken through these development cycles are reflected in the topics covered in Parts Four and Five of the book.

Part Four discusses the various key areas of knowledge required by technology executives to develop a fit-for-purpose and pragmatic, 'sustainable IT strategy'. This section contains a range of interesting and useful topics, including the circular economy, the growth of artificial intelligence (AI), and Generative AI, as well as how to balance the positives of AI with the huge energy consumption required to power these new large language models. I've also included a section called 'Mother Nature' that brings to life the importance of nature in sustaining our planet and the impact that we are having on her. I also examine how to establish a sustainable IT strategy with the proper framework and methodology, as well as how to construct an investment case to mobilise it. The section ends with a review of the new and emerging set of reporting standards that will have significant repercussions for all CIOs and technology leaders in the very near future. I believe the tide has turned, and the largest organisations will no longer have a choice but to take action on decarbonising digital.

People and Culture, the fifth and final section, is arguably one of the most important aspects of digital decarbonisation that needs to be addressed. We will not be able to reduce the carbon impact of digital or make any of our reductions sustainable in their own right if we do not have the skilled people needed to implement the changes required. In this section, I'll discuss topics such as how to set up a successful organisational structure to support decarbonisation, eco-anxiety, and the development of the right culture within an IT organisation in order to assist individuals and teams in their efforts to combat climate change. All research indicates that empowering employees to take action increases their job satisfaction, motivation, and happiness, as well as the value they place on their employers.

I am not interested in simply observing the problem in front of us, and I hope you aren't either. I've written this book to share the information, knowledge, and experience I've gained to help you take action. The book follows my journey of building a world-leading, first-to-market ClimateTech business, and I hope that you can gain from going through the same learning process I went through when creating GoCodeGreen.

Part One: Our Climate has Changed

We Live in a Digital World

Software runs our digital lives and businesses, but we use lots of energy to build it and need hardware, data centres, and networks to keep it running. All of this consumes even more energy. The Information and Communication Technology (ICT) sector created 2.55 billion metric tonnes of CO2e, or 5 percent of global emissions, the same as all the air travel taken in the world in a year (pre-COVID-19), and consumed 6 percent of global energy[7] in 2020. The growth of digital usage and resultant ICT has been estimated at 11 percent annual growth[8], which means that the ICT sector is growing over 12 times faster than the total global population[9].

Data centres are the foundation of the internet. They house the network servers and computers that make the internet possible. They provide companies and personal users with two-way data transfer for internet access and application / data processing, as well as cloud storage for files and programs. They also consume an astonishing amount of electricity because they are packed with technology, such as high-capacity computers that require cooling and maintenance.

There are over 8 million data centres around the world[10]. A single provider, AWS (Amazon Web Services), on its own, covers an estimated 26.6 million square feet[11]. More than 700 of these data centres are considered

---

[7] UK Parliament, 'Energy Consumption of ICT', Sept. 2022

[8] GlobalData, 'Information Technology (IT) Sector ICT Market Size And Forecast (By Country, IT Solution Area And Vertical) To 2027', July 2023

[9] Macrotrends, World Population Growth Rate 1950-2023

[10] DataCenter Knowledge, Apr. 2019

[11] Dgtl Infra, 'Amazon Web Services (AWS) Data Center Locations: Regions and Availability Zones', June 2022

"hyperscale."[12] Hyperscale data centres are BIG, often the size of villages. These vast facilities are full of computers that are intensively using electricity.

It is therefore not surprising that data centres consume approximately 1.3 percent of global electricity, or between 250 and 340TWh[13]. This is equivalent to powering 32 million UK households, and as our use of digital services expands, these figures continue to rise each year.

We have been extremely successful in integrating digital services into our everyday lives. Over 6 hours per day[14], on average, are spent using internet-based services by 64 percent of the world's population[15]. Living in a digital world has significant implications for the amount of energy required to support our increasingly digital existence, and this energy consumption has carbon costs that are growing as the online world continues to gain speed at an alarming rate. As an example, over the last few years, we've witnessed the rise of cryptocurrencies, enormous increases in video streaming, videoconferencing on Zoom and Teams, social media and general internet use, and the explosion of Generative AI.

Even the most basic digital tasks, such as sending an email, require energy, and when you consider that we send over 347 billion emails per day[16], the numbers become enormous. It is estimated that an average email has a carbon footprint of 0.3 gCO_2e[17], which means email-related carbon emissions total 104 tCO_2e a day (or needing 5,000 trees to sequester that amount of carbon), or 38,000 metric tonnes of CO_2e per year (that's the

---

[12] Statista, 'Global Figure of Hyperscale Data Centers 2015-202', Aug. 2023

[13] International Energy Agency (IEA)

[14] Oberlo, Q1 2023

[15] Statista, 'Worldwide digital population 2023', Aug. 2023

[16] Zippia, '75 Incredible email statistics', March 2023

[17] Carbon Literacy Project, 'The carbon cost of an email: update!', Sept. 2022

equivalent of 3,000 UK homes' annual energy usage, or burning 41.6 million tonnes of coal)[18]. Consider that up to 85 percent of all emails sent around the world are spam[19] (unwanted, unsolicited digital communication), and we also start to see the potential for lowering carbon impact through simple actions such as reducing the amount of digital waste.

Let's talk about Bitcoin mining. As you may have read in the media, the carbon footprint of cryptocurrencies is huge. But how big? Bitcoin is estimated to generate 65 million metric tonnes of CO2e a year, which is equivalent to Greece's annual carbon emissions[20]. Think about that for a moment.

---

[18] Calculations based on data from www.pawprint.eco and www.climatelinks.org

[19] According to research completed by Symantec in 2018

[20] The Guardian, 'Bitcoin is terrible for the environment - can it ever go green?', April 2023

The diagram above provides further examples. Streaming music is something many of us do without hesitation or thought on a daily basis. But did you know that a single, viral Spotify song can generate over 3,000 metric tonnes of CO2e annually?[21] Or that streaming a popular Netflix series generates over 90,000 metric tonnes of carbon dioxide equivalent in a year?[22] All of these activities are currently performed without a full understanding of the associated carbon costs. To put this into perspective, we would need to plant a billion trees to sequester Bitcoin's carbon emissions, nearly 200,000 trees to mitigate the carbon footprint of a single viral song, and over 4.4 million trees for that popular television series.

Let's close this section with two final examples of digital carbon impact: Generative AI, the latest trending technology, and one closer to my heart, the estimated digital footprint of the Earthshot Prize.

Generative AI

A spotlight example on Generative AI had to be included, as it is so on trend right now. I'm unashamedly nerdy, so I was super excited to see a recent social post by Aleksa Gordic, X-Google DeepMind, on the technical architecture of OpenAI's ChatGPT-4 infrastructure[23]. I had to perform a few calculations using a GoCodeGreen tool (disclaimer: this is based on assumptions taken from the 'leaked', unverified data) to see if I could estimate the carbon impact cost of it, and the results are shown in the graphic on the next page.

---

[21] WhatsOn, '10 songs causing most CO2 emissions'

[22] The Guardian, 'Streaming's dirty secret: how viewing Netflix top 10 creates vast quantity of CO2', 2021

[23] Find Aleksa's posts on LinkedIn and X, search for 'Aleksa Gordic leaked OpenAI data' from July 2023

OpenAI’s GPT-4 details leaked!

(Aleksa Gordic, X-Google DeepMind)

“COST–

```
* OpenAl's training FLOPS for GPT-4 is ~2.1525, on -25,000 A100s
for 90 to 100 days at about 32% to 36% utilisation.
Part of this extremely low utilisation is due to an absurd number
of failures requiring checkpoints that needed to be restarted.”
```

Estimated tCO2e impact

Assumptions:
25,000 NVIDIA A100 GPUs running 24 hours per day for 100 days at 35% max utilisation, 65% idle time. On Azure HPC, assuming average performance PUE rating and based in California, US (note we take average grid carbon intensity as California is 34% renewable).
No other weightings or assumptions made, pure processor energy utilisation.

c. 8,160 tCO2e = 375,000

That’s correct: 8,160 metric tonnes of CO2e are created based on the energy used by the 25,000 GPU processors during a single training cycle. It would take 375,000 mature trees to absorb that much carbon. That’s a single training cycle run by OpenAl; imagine Google BARD, Microsoft Bing, and Amazon Comprehend all having a comparable impact, and the calculation does not include the subsequent processing of data to generate useable content. We are all aware of how popular Generative AI has become worldwide, and the resulting emissions are becoming astronomical.

The Earthshot Prize

I also turned my attention to the Earthshot Prize. As a nominee for the prestigious environmental prize, I wanted to showcase that even they would be creating a digital carbon footprint that they probably weren't aware of. Based on their own marketing impact report, the diagram below shows the estimated carbon impact of the various digital channels, broadcast content, and social media surrounding the Earthshot process. The Earthshot Prize's marketing and content distribution teams have done an outstanding job of raising awareness of both the award and the need for climate change, but this has an estimated digital carbon footprint of over 1,750 metric tonnes of carbon dioxide. It would require 80,000 trees to absorb this amount of carbon.

The digital carbon impact of the Earthshot Prize

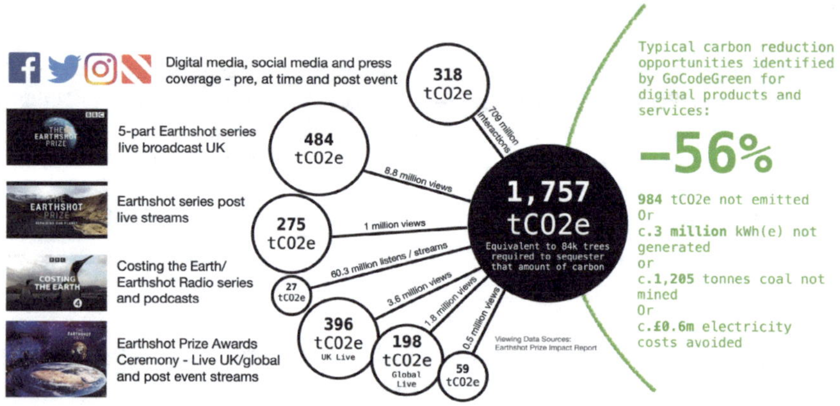

Estimated tCO2e impact of the Earthshot Prize across all digital channels and content - based on viewing metrics from the Earthshot Prize (marketing) Impact Report

c. 1,757 tCO2e = c.80,000 

On the diagram, I've highlighted the quantity of potential carbon reduction (approximately 56 percent) that could be achieved through more energy-efficient software, infrastructure utilisation, and technology choices. I'm hoping to have the opportunity to work with the Earthshot Prize team to validate the calculations and identify the specific steps that could help them reduce their impact.

Summary

It is clear that the scale of emissions related to life in a digital world are immense. We must now work harder to measure and reduce the energy requirements of the entire technology value chain and prevent the release of millions of metric tonnes of CO2e. The opportunity to make more effective use of ICT will not only reduce energy consumption, but it will also help increase the lifespan of the 'embodied carbon' (I'll discuss this in greater detail later) in all the devices and infrastructure components needed for running software.

As Dr. Colette Matthewman, Research and Innovation at the University of East Anglia, commented as part of our Earthshot Prize nomination[24], "GoCodeGreen is really exciting and leading in an otherwise unserved space. What they are doing has huge potential for climate impact, as it's such a widely underpinning area. We believe that by assisting businesses to uncover the hidden carbon emissions in their supply chain, GoCodeGreen can make a real difference where the consumption of energy through technology is concerned."

We have to urgently adopt technology-focused action to reduce energy consumption as quickly as possible. By designing and developing carbon-efficient software, we can help reduce the total energy demand across the entire technology value chain.

We need a global digital reset.

---

[24] Sources: UEA Earthshot prize nominations (uea.ac.uk) and Startups Magazine (www.startupsmagazine.co.uk), March 2023

What is Climate Change?

Long-term changes in temperatures and weather patterns are referred to as climate change. These changes may be the result of natural causes, such as variations in the solar cycle. Since the 1800s, however, human activities have been the primary cause of climate change, specifically the combustion of fossil fuels such as coal, oil, and gas. You need only read, watch, or listen to the news to realise that the effects of climate change are occurring around us every day and are happening now.

The causes of climate change

Rising concentrations of carbon dioxide and other greenhouse gases, such as methane, in the atmosphere produce a 'greenhouse effect' that traps the sun's energy and warms the earth, especially the oceans. Heating the oceans accounts for over nine-tenths of the trapped energy. This greenhouse effect has been known to scientists since the 19th century.

As seen in the diagram below, the Earth becomes warmer as greenhouse gas concentrations in the atmosphere increase. Recent climate change is happening largely as a result of this warming, with smaller contributions from natural influences like variations in the sun's output.

Since before the industrial revolution, carbon dioxide levels have risen by approximately 50 percent[25]. Other greenhouse gases have also increased significantly. This increase in greenhouse gases is almost wholly attributable to human activity. The majority of the increase is due to the use of fossil fuels for energy, the impact of agriculture, and deforestation, and the production of cement, compounds, and metals.

About 45 percent[26] of produced carbon dioxide is released into the atmosphere, while the remainder is absorbed by plants and oceans. Extensive use of land for agriculture, livestock farming, and deforestation reduces the number of trees capable of absorbing carbon dioxide and releases the carbon from these trees back into the atmosphere.

Greenhouse Gas concentrations are at their highest levels in 2 million years

Moreover, emissions continue to rise. As a result, the planet is approximately 1.1°C warmer than it was in the late 1800s. 2011–2020[27] was the warmest decade ever recorded.

Many people believe that climate change primarily entails rising temperatures. However, the rise in temperature is only the beginning of the story. As the Earth is an ecosystem in which everything is interconnected, alterations in one area will influence all other areas.

---

25 Source: climate.nasa.gov, 'Carbon Dioxide Vital Signs' Aug. 2023

26 MIT Climate Portal, 'How much carbon dioxide does the Earth actually absorb?', Jan. 2022

27 IFRF, '2011-2020 declared warmest decade on record', Jan. 2021

Among the current effects of climate change are intense droughts, water scarcity, severe fires, rising sea levels, flooding, melting polar ice, catastrophic cyclones, and a decline in biodiversity.

Every increase in global warming matters

Thousands of scientists and government reviewers agreed in a series of UN reports[28] that limiting global temperature rise to no more than 1.5°C would help us avoid the most severe climate impacts and safeguard a liveable future. Yet based on current national climate plans, global warming is projected to reach around 3.2°C by the end of the century.

The same reports show that emissions that cause climate change originate from every region of the globe and affect everyone, but some countries produce significantly more than others. The 100 countries with the lowest emission rates account for 3 percent of global emissions. The ten countries with the highest emissions contribute 68 percent of the global total. Everyone must take climate action, but those who are primarily responsible for the problem have a greater obligation to do so first.

The chart produced by the IPCC[29] on the next page shows the projected increase in frequency and intensity of high temperatures, which previously only occurred once every ten years on average in a climate without human interference. This calculates that temperature change will accelerate significantly beyond the 1.5°C tipping point; therefore, we must act immediately to avert what will become an irreversible and escalating disaster for our planet.

---

28 United Nations, un.org, 'Climate Action Fast Facts'

29 Source: IPCC, '2021: Summary for Policymakers'

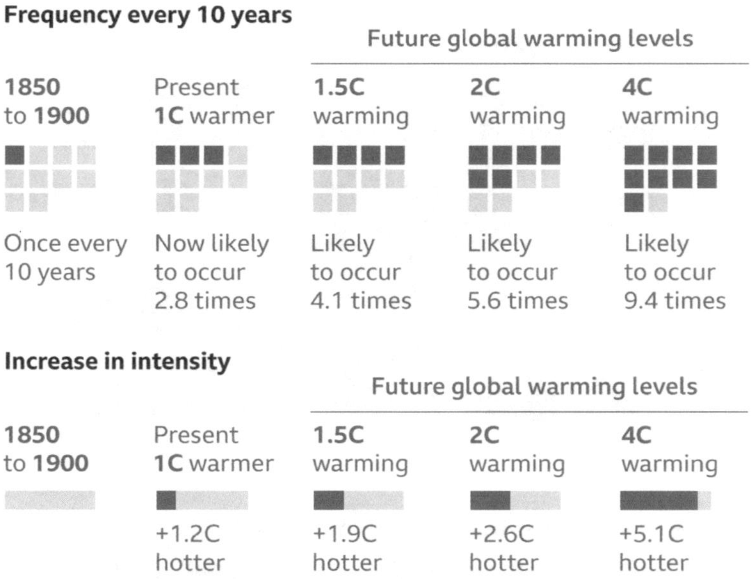

Global climate change is not a problem for the future. Climate changes caused by increased human emissions of heat-trapping greenhouse gases are already having widespread effects on the environment: glaciers and ice sheets are melting, river and lake ice is breaking up earlier, plant and animal geographic ranges are shifting, and natural cycles are being disrupted beyond repair.

Sea ice loss, accelerated sea level rise, and prolonged, more intense heat waves are among the effects of global climate change that scientists have long predicted. Some changes, such as droughts, wildfires, and extreme precipitation are happening more quickly than scientists had previously predicted. The IPCC state that modern humans have never before seen the observed changes in our global climate, and some of these changes are already irreversible over the next hundreds to thousands of years. The diagram below shows the severity of impact if global temperatures rise beyond the 1.5°C and reach 4°C[30].

---

30 Source: World Health Organisation (WHO)

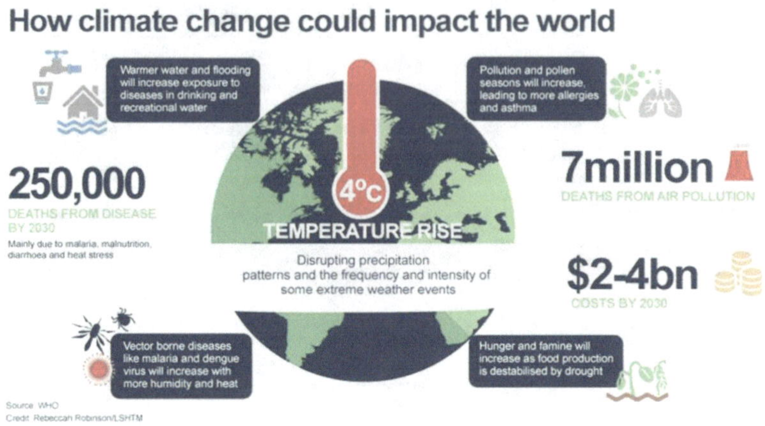

Scientists are clear that global temperatures will continue to rise for many decades, primarily as a result of greenhouse gases generated by human activities. The severity of climate change's effects will therefore depend on the trajectory of future human activities. More greenhouse gas emissions will result in an increase in climate extremes and extensive damage to the planet. However, those future effects depend on the total amount of carbon dioxide we emit, so if we can reduce emissions, we may still be able to avert some of the worst effects.

A call to action: The next few years are critical

In recent IPCC-evaluated scenarios[31], it is clear that limiting warming to 1.5°C (2.7°F) requires global greenhouse gas emissions to peak before 2025 and be reduced by 43 percent by 2030; methane emissions would also need to be reduced by about a third. Even if this was achieved, it seems almost

31 Climate Neutral Group, 'Five future scenarios' Climate Report AR6 IPCC, 2021

inevitable that we will temporarily exceed this temperature threshold, but by the end of the century, we could fall below it if we are able to maintain these levels of reduction. The IPCC asserts emphatically[32]:

> **"It's now or never, if we want to limit global warming to 1.5°C (2.7°F). Without immediate and deep emissions reductions across all sectors, it will be impossible"**

The global temperature will only stabilise when carbon dioxide emissions reach net zero. For 1.5°C (2.7°F), this means achieving net zero carbon dioxide emissions globally in the early 2050s; for 2°C (3.6°F), it is in the early 2070s. It is clear that limiting warming to around 2°C (3.6°F) still requires global greenhouse gas emissions to peak before 2025 at the latest and be reduced by a quarter by 2030.

Without stating the obvious, there is simply no time for observing and talking about the problem any longer. 2025 is only 14 months away at the time of writing these words. Immediate and sustained reduction is needed, and it is needed now.

The next three sections have been written as 'knowledge blocks', intended to quickly give you a good understanding of carbon-based terminology, the Greenhouse Gas Protocol, and the ICT sector, and conclude with my 'recipe for change'.

---

32 IPCC, Jim Skea, Working Group III Co-Chair, April 2022

Knowledge Block One: Understanding Carbon

Carbon emissions & carbon footprint

The unit for measuring carbon emissions, CO_2e, refers to the carbon dioxide equivalent that may be released by a selected human activity. The less CO_2e an activity emits, the lower its environmental impact. For instance, sending a text message produces 0.01g of CO_2e[33], whereas a round-trip flight from London to Hong Kong produces 2.9 metric tonnes of CO_2e[34].

A carbon footprint, on the other hand, is the sum of all CO_2 emissions caused by your actions over a specified time period. Typically, a carbon footprint is measured over the course of one year.

Carbon neutrality

Companies, processes, and products become carbon neutral when they measure and calculate their carbon emissions, take action to reduce those emissions and compensate for any residual emissions that they have produced via carbon-offsetting. Offsetting carbon emissions, in addition to avoidance and reduction, is considered an important step in taking comprehensive climate action.

---

33 BBC.com 'Smart guide to climate change', March 2022

34 SGR.co.uk using BEIS 2021 conversion factors

Greenhouse gases such as CO2 disperse uniformly in the atmosphere, which means that the concentration of greenhouse gases is approximately the same all over the world. Consequently, in terms of the global concentration of greenhouse gases and the greenhouse gas effect, it is irrelevant where on the planet emissions are produced or avoided. Therefore, emissions that cannot be avoided locally can be offset by carbon-offsetting operations elsewhere. This can be accomplished, for instance, through forest conservation, reforestation, or the expansion of renewable energy.

However, offsetting is becoming considered an increasingly unsustainable practice because it bypasses the crucial activity of carbon emission reduction and can lead to global imbalances and inequalities resulting from the impact of offsetting projects (which will be covered in more detail later in this section). There are even those who argue that offsetting can facilitate greenwashing, particularly when the veracity of the carbon offset is unverified.

Carbon removal

Carbon removal, also known as carbon dioxide removal (CDR) or carbon drawdown, is the process of capturing carbon dioxide from the atmosphere and locking it away for decades or centuries in plants, soils, oceans, rocks, saline aquifers, depleted oil wells, or long-lived products like cement. Scientists have proposed many different methods of carbon removal. Some of these are already in use at relatively small scales, whereas others remain in the early stages of research and development. Technologies and practices for implementing carbon removal are often called Negative Emission Technologies (NETs).

Carbon removal matters because 45 percent of the carbon dioxide that humanity emits will remain in the atmosphere for up to a thousand years, with roughly 10–25 percent of it persisting for tens of thousands of years. Removing and sequestering that carbon dioxide could permanently reduce climate risk by slowing or even reversing climate change.

Versus carbon capture usage and storage

Carbon removal is not the same as Carbon Capture, Usage and Storage (CCUS). There are two variants of CCUS, Carbon Capture and Storage (CCS) captures carbon dioxide from a smokestack or flue, such as in a coal-fired power plant or a cement factory, and then sequesters that carbon dioxide underground. Whereas, processes that capture carbon dioxide and use it to produce commercial products are known as Carbon Capture and Use.

Carbon offsetting

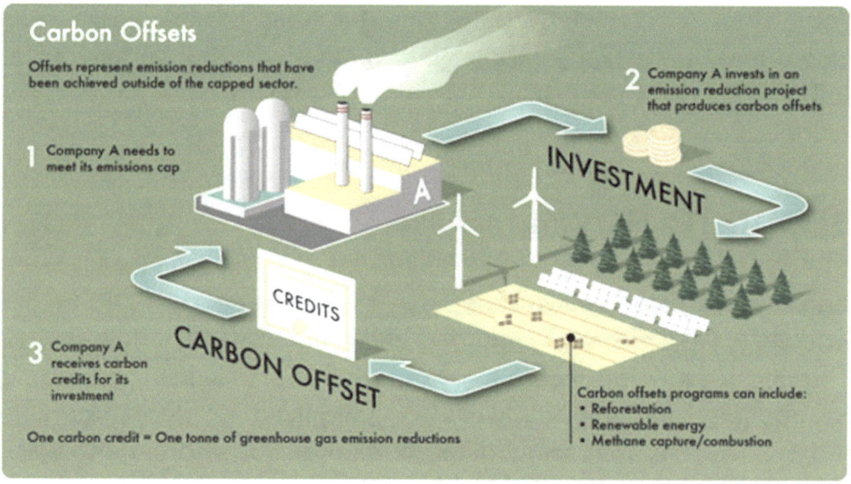

Carbon offset schemes allow individuals and companies to invest in environmental projects around the world in order to balance their own carbon footprints. The projects are usually based in developing countries and are most commonly designed to reduce <u>future</u> emissions. This might involve rolling out clean energy technologies or purchasing and ripping up carbon credits from an emissions trading scheme. Other schemes work by soaking up CO2 directly from the air through the planting of trees. The diagram[35] on the previous page provides a good illustration of the different types of projects available.

Some people and organisations offset their entire carbon footprint, while others aim to neutralise the impact of a specific activity, such as taking a flight. To do this, the holidaymaker or business person visits an offset website, uses the online tools to calculate the emissions of their trip, and then pays the offset company to reduce emissions elsewhere in the world by the same amount, thereby rendering the flight 'carbon neutral'.

Costs for offset schemes differ widely, but a typical fee for each tonne of CO2 offset would be around £11 or $13 per tonne[36]. At this price, a typical British family would pay approximately £45 / $55 to offset a year's worth of gas and electricity consumption, whereas a round-trip flight from London to San Francisco would cost approximately £20 / $24 per ticket.

Carbon offsetting can only be effective if it reduces carbon emissions more than they would have been without it and if it is permanent. Renewable energy, energy efficiency, carbon sequestration, and aviation offsets are useful first steps, but reducing emissions at their source is most effective and more sustainable.

A form of neocolonialism?

Numerous businesses have made commitments to achieve Net-Zero emissions. In order to accomplish this challenging goal, an increasing number of them are turning to carbon credits (a measurement to cap

---

35 Source: MeetGreen, meetgreen.com, 'Carbon Offsets - The Basics'

36 Financial Times, 'Cost of neutralising carbon emissions soars as demand escalates', per data from S&P Global Platts, Jan. 2022

permitted emissions). McKinsey[37] estimates that by 2030, the voluntary carbon credit market could be worth more than $50 billion globally.

In contrast, carbon offset refers to the reduction, avoidance, or sequestration of carbon dioxide equivalents in order to offset emissions that are taking place in another location. The majority of carbon offset initiatives involve reforestation. However, reducing carbon emissions by simply planting more trees in developing countries as part of a carbon offset project can easily turn into another form of neocolonialism.

Neocolonialism is the practice of exerting influence over a country by way of economics, globalisation, cultural imperialism, and conditional aid. Climate colonialism, a new form of neocolonialism, is the result, according to the Georgetown philosopher Olfmi Táíwò[38], who argues that these initiatives to slow down the rate of climate breakdown create a dominance for powerful countries over weaker nations and people. There remain too many situations where affluent nations are making investments in the Global South and imposing maintenance rules on forests and land inhabited by indigenous populations that date back thousands of years.

Today, trees are among the most common carbon-offsetting methods. The voluntary offset market witnessed the highest proportion of transactions involving forestry and land use offsets in 2019 (56.4 percent). Renewable energy and domestic device projects accounted for the remaining 21.3 and 8.8 percent, respectively[39]. Advocates of carbon offsetting through forestry frequently characterise such initiatives as 'win-win-win' because they serve to mitigate climate change, safeguard forests, and provide avenues for

---

[37] McKinsey Sustainability, 'A blueprint for scaling voluntary carbon markets to meet the climate challenge', Jan. 2021

[38] psmag.com 'How the green new deal can avoid climate colonialism', Feb. 2019

[39] thehill.com 'Avoiding carbon colonialism: developing countries can't pay the price for pollution', April 2021

community development. In reality, carbon offsetting projects can fail catastrophically on all three counts.

Developing regions are incentivised to privatise and commercialise their forest and land resources by the global carbon offset market. Developed country governments, organisations, corporations, and individuals purchase offsets, whereas developing country suppliers provide the offsets. Currently, the countries with the highest quantity of purchasers of voluntary carbon credits are Switzerland, the United States, France, the United Kingdom, and Germany. The countries that produce the most offsets are Cambodia, Peru, Brazil, Kenya, Zimbabwe, Bolivia, and Indonesia[40]. One could easily view forest and land-use-based carbon offsetting as a means of appropriating land in the Global South in the alleged pursuit of addressing climate change. In this current model, the Global North continues to pollute while communities in the Global South lose land and sovereignty.

The process of establishing carbon offsetting sites often entails the deliberate removal of indigenous communities from areas and resources that were previously under communal or public ownership. As an illustration, according to a 2011 Oxfam study[41], the reformation of Ugandan timber plantations by the United Kingdom-based New Forests Company for the purpose of producing carbon offsets displaced at least 22,500 people from their homes. Some experts have described such displacement as a violation of human rights. In other instances, offsetting projects have facilitated the acquisition of indigenous land by a foreign company or state.

So the reality is that the Global North makes the majority of purchases, whereas the majority of carbon offset initiatives are located in the Global South. People in industrialised nations are using carbon offsets almost as a convenience instead of holding their governments accountable and altering

---

40 'Fertile Ground - state of forest carbon finance 2017', Hamrick and Gallant, Dec. 2017

41 Oxfam case study: 'The new forests company and its Uganda plantations', Grainger and Geary, Sept. 2011

their production and consumption patterns. They demand the Global South alter their economic patterns in order to compensate for emissions from the Global North, while vulnerable developing nations endure the effects of climate change. Next time you hear mention of carbon offsets, be cautious. It doesn't mean that they should be stopped, but they must be more strongly regulated in the future.

Decarbonisation

Decarbonisation is the process of reducing CO2 emissions caused by human activity, with the ultimate aim of eliminating them. The Paris Agreement of 2015 established the objective of limiting global warming to well below 2°C above pre-industrial levels and pursuing activities to limit it to 1.5°C, in part by pursuing carbon neutrality by 2050. To achieve this, we require a significant reduction of global greenhouse gas emissions (including CO2) to limit global temperature increases.

In order to achieve Net-Zero emissions in practice, fossil fuels must be replaced with alternative, low-carbon energy sources in conjunction with activities to reduce the demand for energy and products and services requiring the continued extraction of raw materials and fossil fuels from the ground.

Decarbonisation will only be realised by reducing CO2 emissions across all industries. Utilising alternative energy sources based on green electricity and green molecules (such as biofuels and hydrogen) will be essential to achieving decarbonisation.

Embodied carbon

Embodied carbon refers to all CO2 emitted in the production of materials. It is estimated based on the energy used to extract and transport materials as well as the emissions generated from manufacturing processes. For example, the embodied carbon of a building can include all the emissions from the construction materials, the building process, and all the fixtures and fittings inside, as well as from deconstructing and disposing of it at the end of its lifetime.

Because of the length and complexity of the supply chain, embodied carbon is particularly difficult to calculate in the ICT industry. Some of the raw materials are extremely rare and sourced from all over the globe. For example, rare earth metals mined in Brazil are being transported to component manufacturers, who are frequently located in the Far East. The manufactured components are then sent to the assembly facilities, which may be located in Europe or the United States. The final product is then delivered to the consumer via extensive distribution channels.

The following diagram[42] shows the pathway for embodied carbon using the construction and building industries though similarities do exist with ICT and the process flow is almost identical. Technology has a complex and difficult-to-calculate chain due to the sheer range and variety of raw materials and manufacturing processes involved; however, progress is starting to be made on better measuring embodied emissions, which will enable us to make more informed product selection decisions.

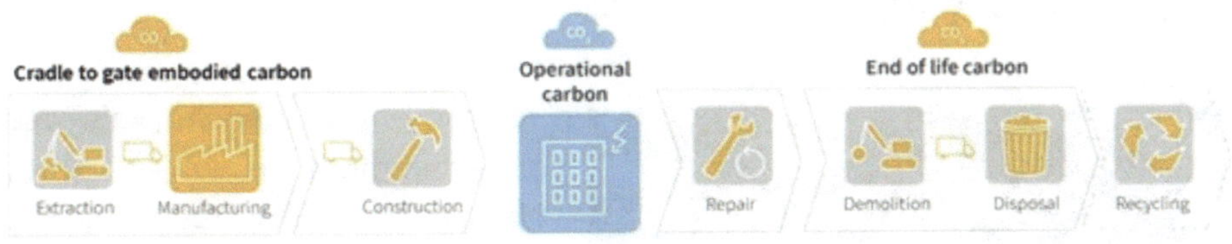

However, as this clarity improves and evolves, it poses a challenge to the accuracy of reporting. Different manufacturers are publishing different

42 Source: U.S General Services Administration, 'Embodied carbon: Next steps for GSA', July 2021, copyright Carbon Leadership Forum

embodied carbon figures for comparable pieces of equipment. This might be the result of presumptions about where component manufacturing and product assembly take place, and the Life Cycle Analysis (LCA) methodology they employ may also have an impact. Sometimes they also make different decisions on how to define the product (for a server, it could include differing categorisations for the number of CPUs, grade and size of memory, or type of storage), and also how they define pre-use (i.e., whether they set a boundary at the distributor or extend this to a retail outlet, for instance).

Despite the issues around accuracy, it is important to recognise that every product and material has an embodied carbon cost. This will rise if it is manufactured in an area where a lot of fossil fuels are used in electricity generation and if there is a lot of transport required to deliver the product.

The technology sector is one of the most significant contributors to carbon emissions. For instance, the annual carbon footprint of mobile phone production (146 million metric tonnes of CO2e) is equivalent to that of the Philippines (a country with over 100 million inhabitants)[43]. However, by extending the lifespan of technology hardware through refurbishment, reuse, repair, and ultimately recycling, we can reduce the global consumption of technology and the industry's carbon footprint as a whole.

Apple and Mother Nature

You may have seen Apple's recent advertisement in which Mother Nature pays Tim Cook and his team a visit to get an update on their environmental performance. Further in the book, I've added a 'Mother Nature' section that provides a broader perspective on

43 therestartproject.org 'Mobiles: the global carbon footprint'

the significance of nature in addressing the climate crisis, and then you can decide if she would be pleased with humanity's scorecard.

On viewing the advertisement, I wanted to try to work out a few facts and figures that might bring the concept of embodied carbon to life for the reader and also demonstrate the impact we are having as a result of the digital lives we lead. There is no greater symbol of living in our digital world than a smartphone, and I apologise to Apple for singling you out, but your ad did its job and caught my attention.

An iPhone 14 Pro weighs 240 grammes. Apple[44] estimates that its embedded emissions (emissions deriving from production, transportation, and disposal) are as high as 116 kg CO2e. The average embodied carbon across all iPhone 14 models is 84 kg CO2e. That's 350 times its own weight in embodied carbon costs.

According to Omdia[45], Apple sold 64 million iPhone 14s (in all model variations) in the first half of 2023. Let's assume that they sell 100 million devices in the full year (for sales figures, see the Statista[46] report). That equates to 8.4 million metric tonnes of CO2e per year for iPhone 14 production and sales.

The Energy Savings Trust[47] reports that the average UK household with a family of four emits 8.8 metric tonnes of CO2e annually. Using this figure, the embodied carbon of iPhone 14s sold globally in 2023 would be equivalent to the total emissions of 955,000 UK households for a full year (or the emissions of 3.82 million people).

---

44 apple.com, 'Product Environmental Report: iPhone 14 Pro', Sept. 2022

45 Source: Cult of Mac, referencing Omdia, Aug. 2023

46 Statista, 'Apple worldwide shipments of smartphones from 2010 to 2023, by quarter', Aug. 2023

47 energysavingstrust.org.uk, Oct. 2020

If up to 80 percent of the carbon caused by those iPhone 14s is sequestered by nature via land and oceans, then the remaining 20 percent, or 1.7 million metric tonnes of CO2e, remains in the atmosphere. And that is only if we allow Mother Nature to go about her job without disturbing her natural ecosystems and reducing her ability to soak up the CO2e we have produced as a result of those iPhone 14s.

That is 1.7 million metric tonnes of excess carbon dioxide produced by humans that might remain in the atmosphere for many thousands of years. All these excesses add up, and that is why the planet's climate is in a state of crisis. This is one example of a single smartphone model from one manufacturer. Think about that[48].

Stop Press: On September 25, 2023, Apple announced their latest iPhone, the 15 Pro. Made with an outer shell of titanium. Mother Nature should consider the following in her next status report for Apple: The average embodied carbon across models has increased from 84kg CO2e to 87kg CO2e (although the pro version is reduced from 116kg CO2e for the iPhone 14 Pro to 107kg CO2e)[49].

---

48 To Emily, a 13-year-old future vet, who dared to challenge. I added detail to this section just for you. Read, then go tell those other kids at school (and your teacher) that you are right. Digital and technology has a carbon footprint and does impact our planet.

49 apple.com, 'Product Environmental Report: iPhone 15 Pro and Pro Max', Sept. 2023

The mining and processing of 'virgin' titanium (i.e., not recycled) have environmental impacts, including soil erosion, water contamination, and habitat devastation, all affecting the natural ecological balance. According to a study by Marine Georesources and Geotechnology[50] in Kenya, "the mining process involves the removal of vegetation cover and top soil, which leads to severe erosion with an eventual serious impact on the population downstream. Toxic chemicals used in heavy mineral separation processes and disturbance or redistribution of sediment spell a disaster for coastal waters." I can already see Mother Nature shaking her head in disapproval.

The more positive news is that titanium is 100 percent recyclable, reducing the need for new mining and processing activities in the future. Extending the lifespan of these new iPhone 15s and implementing a major incentive scheme to encourage the circular and reuse of these devices could help to better manage their long-term impact on the climate.

I commend Apple's investment in producing device-level environmental reports; after all, measurement and increased transparency help us make better decisions and permit review and critique. I'd much rather have that than the deadly silence from some manufacturers.

You know that feeling, usually around the two and a half-year mark, where the apps start getting slower on your smartphone and you start thinking about replacing it? What I'd love Apple to do next is push every application in their App Store through the GoCodeGreen assessment process and verify that action is being taken to make the software that runs on their devices as energy efficient as possible. This would contribute to extending the lifespan of the embodied carbon in every iPhone on the planet. If we were part of the App Store Review process, that would help change the sustainability impact that Apple could make overnight. I think Mother Nature would approve.

---

50 Marine Georesources and Geotechnology, 'Environmental Impact Assessment of the Proposed Titanium Mining Project in Kwale District, Kenya', J.O.Z Abuodha, Nov. 2010

Net-Zero

"A target of completely negating the amount of greenhouse gases produced by human activity is to be achieved by reducing emissions and implementing methods of absorbing carbon dioxide from the atmosphere."

Oxford Languages

Simply put, 'Net-Zero' refers to reducing greenhouse gas emissions to as close to zero as possible while allowing oceans and forests to absorb any remaining emissions.

Transitioning to a Net-Zero world is one of humanity's greatest challenges. It requires nothing less than a complete transformation of production, consumption, and transportation.

The energy sector is responsible for approximately three-quarters of current greenhouse gas emissions and holds the key to mitigating the most severe consequences of climate change. Substituting carbon-emitting coal, gas, and oil-fired energy with wind and solar power would significantly reduce carbon emissions.

The Science Based Targets Initiative (SBTi) has defined a Corporate Net-Zero Standard[51] and is the world's only framework for corporate Net-Zero target setting in line with climate science. It includes the guidance, criteria, and recommendations companies need to set science-based Net-Zero targets consistent with limiting global temperature rise to 1.5°C. The corporate Net-Zero standard has the following core components:

---

[51] See: https://sciencebasedtargets.org/resources/files/Net-Zero-Standard.pdf

- To prioritise rapid, deep emission reductions by setting near-term targets: Rapid, deep cuts to direct and indirect value-chain emissions are the most effective and scientifically sound way of limiting global temperature rise to 1.5°C. Halving emissions by 2030 must be the overarching priority for all companies.
- Set long-term targets: The Net-Zero Standard also requires companies to set long-term science-based targets to cut all possible emissions before 2050. Most companies must reduce emissions by more than 90 percent.
- To neutralise residual emissions: After a company has achieved its long-term target and cut emissions by >90 percent, it must use permanent carbon removal and storage to counterbalance the final <10 percent of residual emissions that cannot be eliminated. A company is only considered to have reached Net-Zero when it has achieved its long-term science-based target and neutralised any residual emissions.
- Climate finance beyond the value chain: The SBTi wants to incentivise and catalyse as much climate finance as possible from the private sector. There is an urgent need to scale up near-term climate finance to achieve Net-Zero before 2050. Companies should follow the mitigation hierarchy above and make these investments in addition to deep emissions cuts, not instead of them.

The following graphic summarises the standard and what it means for any organisation seeking to commit to and achieve the Net-Zero target.

Net Zero is measured on a different scale and dimension.	"The boundary of a Net Zero target includes global scope 1, 2 and 3 emissions of the organisation" **Carbon Trust**
No organisation can achieve Net Zero unless they also measure their digital and ICT value chain.	"The Net-Zero Standard covers a company's entire value chain emissions, including those produced by their own processes (scope 1), purchased electricity and heat (scope 2), and generated by suppliers and end-users (scope 3)." **Science Based Targets**
Decarbonisation first. Offsetting limited to 5-10% only. Cover all GHG emissions, not just CO2.	" ... companies will require deep decarbonisation of 90-95% to reach net-zero under the Standard ... (and neutralise) any residual emissions at the net-zero target year and any GHG emissions released into the atmosphere thereafter." **Science Based Targets**

SOURCE: www.sciencebasedtargets.org/resources/files/Net-Zero-Standard.pdf

The challenge we all face is that there is a significant difference between what we are currently emitting (11 percent growth every year) and the amount of reduction that will be necessary by 2030 (43 percent) if we want to avoid reaching the tipping point, the 1.5°C temperature increase that will allow us to avoid the worst climate impacts. The objective of Net-Zero is to keep the rise in global average temperatures within certain limits, and these limits are defined by a finite budget of carbon dioxide that is allowed into the atmosphere alongside other greenhouse gases. Beyond this 'global carbon budget', any further release must be balanced by removal.

Global Carbon Budget

"We have less than 10 years before we exhaust our global carbon budget."

Intergovernmental Panel on Climate Change (IPCC)[52]

Since 2005, a group of researchers has published the "Global Carbon Budget," a report that quantifies carbon dioxide (CO2) emissions for the previous year. The report calculates the amount of CO2 emitted by human activities and stored in carbon reservoirs on land and in the ocean.

A budget for carbon?

According to the World Meteorological Organisation[53], to monitor a financial budget, one must know the balances between income and expenditure. Likewise for carbon, with CO2 in the atmosphere as the relevant 'account', one quantifies inputs (sources) and outputs (sinks). A global carbon budget determines the input of CO2 to the atmosphere by emissions from human activities, balanced by output (storage) in the carbon reservoirs on land or in the ocean. The carbon budget can be written as a relatively simple equation:

---

52 IPCC, Feb. 2022

53 public.wmo.int 'The Annual Global Carbon Budget', 2017

Sources = amounts in the atmosphere plus amounts in the sinks, where sinks represent removals from the atmosphere but inputs to the ocean or land.

For any given amount of emissions, if less carbon goes into the land or ocean, more will remain in the atmosphere. More specifically:

Emissions from fossil fuel combustion + emissions from land use change = growth in the concentration in the atmosphere + amount going (chemically and biologically) into the ocean + amount going into terrestrial vegetation and soils.

According to the IPCC, the global carbon budget calculated in 2020 gave us a budget of 500 billion metric tonnes of CO2e before warming exceeds the 1.5°C tipping point. In 2021, the IPCC recalculated the outstanding budget to be 400 billion metric tonnes of CO2e, and by the end of 2022, we had already emitted and consumed the budget by a further 80–100 billion metric tonnes of CO2e. For an average country, its share of this global CO2 budget will run out in 8 years. For a high-emission country such as the UK, its share of the carbon budget will run out in two years[54]. To stay within 1.5°C of global warming, immediate radical action is needed to phase out the burning of fossil fuels.

I've included the full-page infographic produced by the Global Carbon Project[55] below. It clearly shows that we have the capacity to reduce emissions, as indicated by the record reductions experienced during the COVID-19 pandemic. It also highlights our capacity to continue emitting at high rates in the rapid rebound post-COVID-19 and the scale of emissions (1.4 billion metric tonnes of CO2e each year) that are required in order for us to manage the global carbon budget.

---

[54] carbonindependent.org 'The global carbon budget will run out in 8 years', Sept. 2023

[55] globalcarbonproject.org, acknowledging Natalie Porter at the UEA for creating the excellent poster shown on the next page.

Global Carbon Budget 2021

CO_2 emissions rebound towards pre-COVID levels

Global fossil CO_2 emissions in 2021 are set to rebound 4.9% after a record 5.4% drop in 2020.
This follows a decade of strong and growing energy decarbonisation which reduced the growth of emissions.

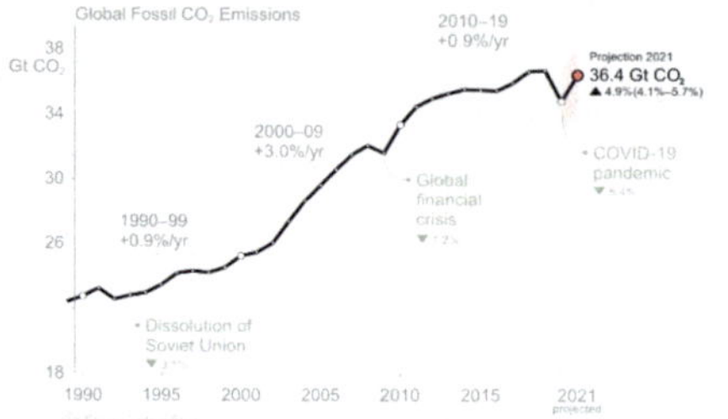

CO_2 emissions cuts of 1.4 billion tonnes are needed each year on average to reach net zero by 2050

Coal and gas use grow more in 2021 than they fell in 2020.
Oil use remains below 2019 levels.

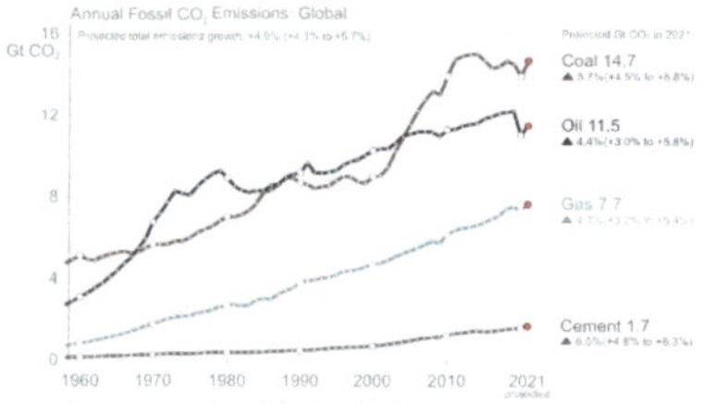

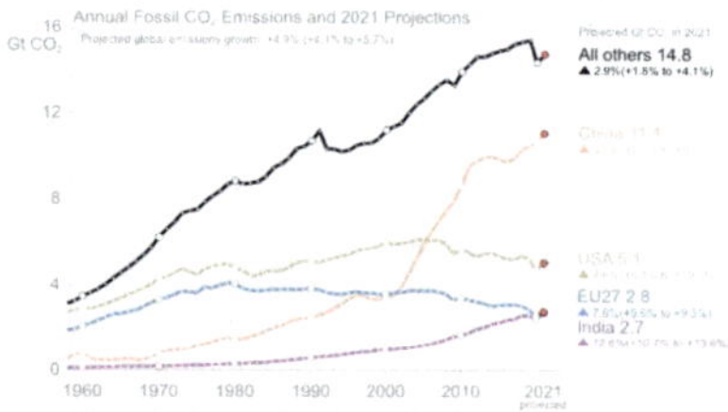

Emissions return to pre-COVID trends in USA, EU27 & India but further growth was sparked in China.

The level of CO_2 continues to increase in the atmosphere, causing climate change.
Emissions from deforestation and other land-use change remain high, partly offset by removals from regrowth of forest and soil recovery.

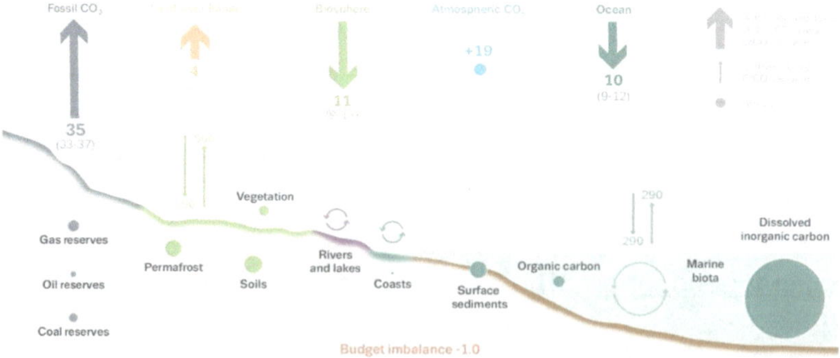

Knowledge Block Two: Understanding the Greenhouse Gas Protocol

The Paris Agreement

The Paris Climate Agreement is an international treaty that commits most of the world's governments to addressing climate change. The Paris Agreement is the world's first comprehensive climate treaty and has the goal of stopping the world's average temperature from rising by more than two degrees, and ideally by less than 1.5°C.

This would likely prevent the worst effects of climate change (although it would still cause significant damage, particularly to those who contributed the least to this crisis). But at the moment, the world isn't yet on track to achieve that goal.

Nearly every nation in the world has signed the Paris Agreement. Iran, Turkey, Eritrea, Iraq, South Sudan, Libya, and Yemen are the only ones who have not joined.

Is progress being made?

Many experts agree that it has helped to accelerate global climate action, but not by enough yet. Recent studies have shown that the commitments and pledges made over the last ten years are 'within reach', but also that reaching them will be difficult. To get back on track, nations need to set more ambitious short-term goals for reducing their emissions. They also need to exert greater effort to fulfil their promises. See the diagram below showing current progress by country[56]. There is a frightening lack of green, and there remain very few role models to help lead the way.

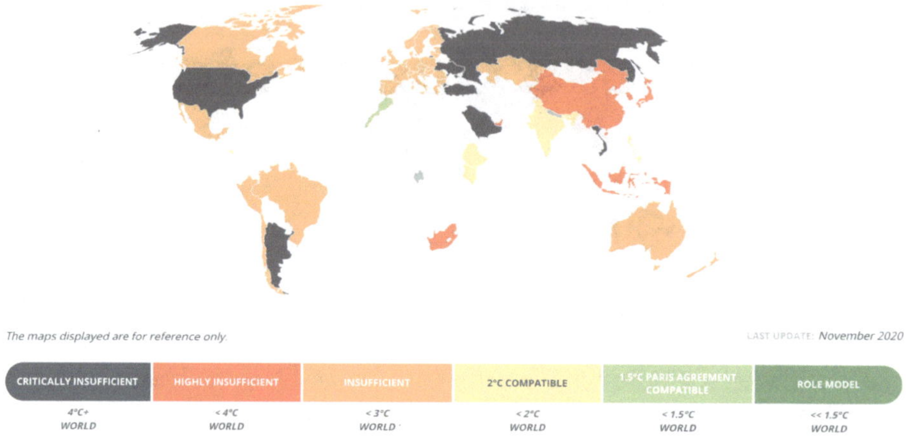

The agreement holds that global commitments should be sufficient to keep global warming below 1.5°C, the threshold for irreversible and harmful change. At the heart of the Paris Agreement are NDCs (Nationally Determined Contributions), where each government can decide how big its reduction will be, which can clearly lead to imbalanced contributions around the world. Many experts are clear that the G20 nations, responsible for more than 75 percent of global GHG emissions, must show greater leadership through their contributions and actions if we are to achieve the Paris Agreement global goals.

The Paris Agreement works on a 'ratchet mechanism,' a cycle of increasing ambition through revised NDCs submitted every five years. The number of new or updated NDCs submitted to the United Nations Framework Convention on Climate Change (UNFCCC) grew from two in 2019 to 143 submissions in 2021 as countries worked hard to get Paris Agreement

---

56 Map sourced from Greenpeace, 'What is the Paris climate agreement and why does it matter', March 2021

timelines and processes back on track after the COVID-19 pandemic. Despite this increase in NDCs, GHG emissions are projected to only decrease by 9 percent in 2030 compared to 2010 levels. But while heading in the right direction, it is still far short of the 45 percent slash in GHG emissions by 2030 that is required to reach net zero emissions by 2050[57].

Developed nations have a responsibility under the Paris Agreement to deliver $100 billion a year in climate finance by 2030. But new data from Oxfam and the Organisation for Economic Co-operation and Development (OECD) shows that wealthier nations are falling short and will continue to miss their target, meaning that climate-vulnerable countries will face a shortfall of between $68 billion and $75 billion in total over a six-year target period. Meanwhile, an astounding $423 billion is being spent annually to subsidise fossil fuels[58]. This is four times the amount being called for to help poor countries tackle the climate crisis.

Is progress being made? Yes. But it is too slow, and the commitments to support action by developed nations remain below the levels needed to make the scale of change needed. At present, if every country hit its current Paris Agreement target, global emissions would only be 1 percent lower by 2030, and the world would warm by 2.1–3.3°C.

The Greenhouse Gas Protocol (GHGP)

I've shown you how the Paris Agreement commits countries to reduce greenhouse gas emissions in order to keep the global temperature rise below 1.5°C, thereby avoiding the most severe effects of climate change. The GHG Protocol arose from the need to assist countries and companies in

---

57 United Nations Development Programme, 'The state of climate ambition', 2021

58 Organisation for Economic Cooperation and Development (OECD) (2021) and Oxfam International (2021)

accounting for, reporting, and mitigating emissions and to create an action agenda to address climate change based on a standardised measurement of GHG emissions.

The Greenhouse Gas (GHG) Protocol is a multi-stakeholder partnership formed by the World Resources Institute (WRI) and the World Business Council for Sustainable Development (WBCSD) with the participation of businesses, non-governmental organisations (NGOs), and governments. Launched in 1998, the GHG Protocol's mission is to develop internationally accepted greenhouse gas (GHG) accounting and reporting standards and tools and to promote their adoption so as to achieve a global economy with lower emissions. It has established a comprehensive, global, standardised framework for measuring and managing emissions from private and public sector operations, value chains, products, cities, and policies in order to facilitate greenhouse gas reductions across the board.

Greenhouse Gas (GHG)

Greenhouse gases (also known as GHGs) are gases that capture heat in the earth's atmosphere. Since the Industrial Revolution, humans have increased the amount of greenhouse gases they release into the atmosphere, and the quantity has accelerated dramatically over the past century, contributing to global warming. In the past 30 years, the rise in global temperature has now been the highest ever recorded.

GWP and GHGs

Greenhouse gases other than carbon dioxide contribute to global warming. Each greenhouse gas (GHG) has a different Global Warming Potential (GWP) and stays in the atmosphere for a varying amount of time.

GWP is a measure of the amount of energy the emissions of one metric tonne of a gas will absorb in comparison to one metric tonne of carbon dioxide (CO2) over a specified time period. The greater the GWP, the greater the Earth's warming induced by a particular gas relative to CO2 during that time period. The time period usually used for GWPs is 100 years. By establishing a standardised unit of measurement, GWPs enable the aggregation of emissions estimates for various gases and allow policymakers to compare opportunities to reduce emissions across sectors and gases.

The diagram below provides a view of the Global Warming Potential (GWP) and scale of each major GHG emission[59].

HOW GREENHOUSE GASES WARM OUR PLANET

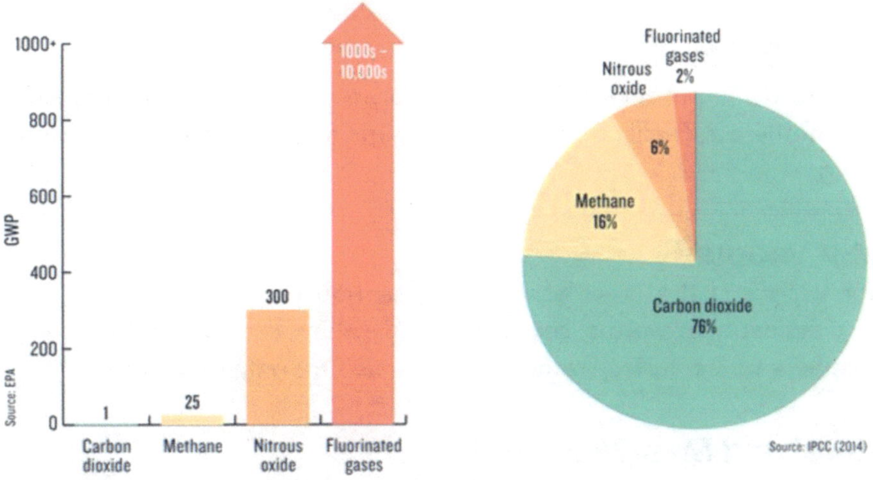

The global warming potential (GWP) of human-generated greenhouse gases is a measure of how much heat each gas traps in the atmosphere, relative to carbon dioxide.

How much each human-caused greenhouse gas contributes to total emissions around the globe.

---

[59] Source: Natural Resources Defense Council), nrdc.org, 'Greenhouse Effect 101', June 2023

The other forms of greenhouse gases[60] are:

Carbon Dioxide (CO_2)

Natural processes such as volcanic eruptions, plant respiration, and animal and human respiration all produce CO_2. Since the Industrial Revolution began in the 1800s, the concentration of CO_2 in the atmosphere has increased by 47 percent due to human activities such as the combustion of fossil fuels and widespread deforestation. Due to the abundance of CO_2 in the atmosphere, it is the main contributor to climate change.

Methane

Decomposition is a natural process that produces this greenhouse gas. Again, however, human activity has disrupted the natural equilibrium. The production of oil and gas, landfill waste dumps, rice farming, and cattle farming all emit significant amounts of methane.

Nitrous Oxide

Nitrous oxide is produced through the large-scale use of commercial and organic fertilisers, fossil fuel combustion, nitric acid production, and biomass burning.

Water Vapour

Water vapour is the most abundant greenhouse gas. It increases as the earth's atmosphere warms, but unlike CO_2, which can remain in the earth's atmosphere for centuries, water vapour persists for only a few days.

Natural and Man-Made GHGs

This group of gases is generated naturally, but their increasing concentration in the atmosphere is a result of human activity. In contrast, the three industrial fluorinated gases–hydrofluorocarbons (HFC), perfluorocarbons (PFC), and sulphur hexafluoride (SF_6) – are exclusively produced by industrial processes

[60] Source: carboncollective.co, 'Carbon Dioxide Equivalent', Sept. 2023

and do not occur naturally. Even though they are present in extremely low concentrations in the atmosphere, they are extremely potent heat trappers. The GWP of SF_6, which is used in high-voltage electrical equipment, is approximately 23,800 times greater than that of CO_2[61].

Carbon Dioxide Equivalent (CO2e)

CO2e is referenced throughout the book, but what does it mean? Carbon dioxide equivalent (CO2e) is a metric unit used to compare greenhouse gas emissions based on their global warming potential (GWP).

CO2 (carbon dioxide) is an odourless, colourless greenhouse gas emitted when fossil fuels (such as natural gas, oil, and coal) are burned. However, CO2e, or carbon dioxide equivalent, is more accurate, as it includes all the molecules that will absorb heat and warm our atmosphere.

The objective is to compare the impact of each greenhouse gas to the amount of CO2 required to generate the same amount of heat. As a result, a carbon footprint, which consists of numerous greenhouse gases, can therefore be expressed as a single number.

GHG Protocol product lifecycle assessments

As climate change awareness and concerns develop, there is a growing demand for greater environmental transparency, clarity, and accountability. Companies are increasingly being asked to measure and disclose their GHG inventories, which include emissions from their own products and their

---

[61] Source: National Atmospheric Emissions Inventory, https://naei.beis.gov.uk/

supply chain. To ensure long-term success in a competitive business environment and to be ready for any future product-related policies and programmes, businesses must be able to understand and manage the GHG risks associated with their products.

The GHG Protocol Product Standard was developed to provide standards and guidance to assist businesses in achieving these goals. The standard enables the following:

- An approach to GHG accounting from a value chain or life cycle perspective.
- It helps account for life cycle emissions at the individual product level.
- It enables a company to target individual products with the greatest potential for reductions.
- It helps a company identify GHG reduction opportunities, track performance, and engage suppliers at a product level.

GHG Protocol Scopes 1, 2 and 3

In order to take action to reduce emissions, it is necessary to understand and measure where they originate from in the first place. The three scopes are a way of categorising the different kinds of emissions a company creates in its own operations and in its wider 'value chain' (its suppliers and customers).

According to the GHG Protocol, the primary purpose of this categorisation is to "help delineate direct and indirect emission sources" and "ensure that two or more companies will not account for emissions within the same scope."

According to the Greenhouse Gas Protocol[62], "Developing a full [greenhouse gas] emissions inventory, including Scope 1, Scope 2, and Scope 3 emissions, enables companies to understand their full value chain emissions and concentrate their efforts on the greatest opportunities for reduction."

---

62 GHG Protocol, ghgprotocol.org

Scope 1 and 2 emissions are those that a company owns or controls, whereas scope 3 emissions are a result of the company's activities but come from sources that it does not own or control. The diagram below provides an overview of GHG Protocol scores and emissions across the value chain[63].

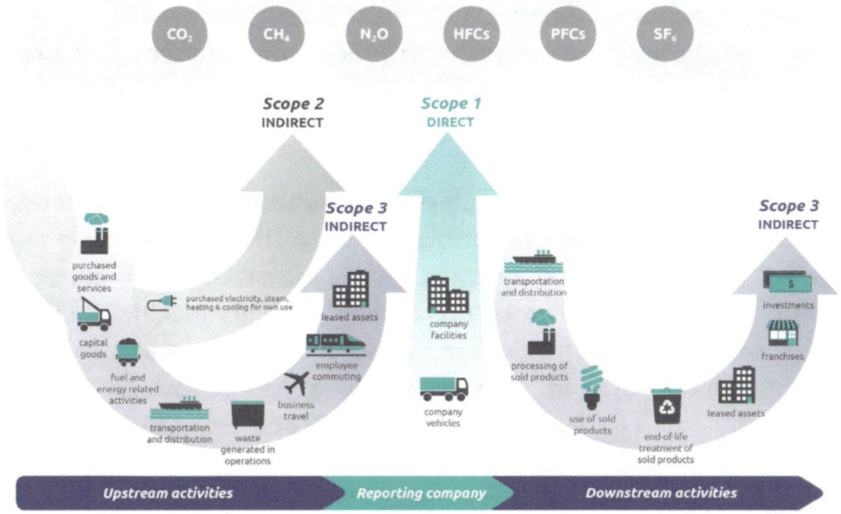

Scope 1 Emissions

Scope 1 emissions include direct emissions from the company's owned or controlled sources. This includes on-site energy like natural gas and fuel, refrigerants, and emissions from combustion in owned or controlled boilers and furnaces, as well as emissions from fleet vehicles (e.g., cars, vans, trucks, and even helicopters for hospitals). Scope 1 emissions also encompass

63 Source: WRI/WBCSD 'Corporate Value Chain (Scope 3) Accounting and Reporting Standard

emissions that are released during industrial processes and on-site manufacturing (e.g., factory fumes and chemicals).

Scope 2 Emissions

Scope 2 emissions include indirect greenhouse gas emissions from purchased or acquired energy, like electricity, steam, heat, or cooling, generated offsite and consumed by the reporting company. For example, electricity purchased from a utility company is generated offsite, so it is considered an indirect emission.

If the reporting company, for example, an industrial facility, generates its energy on-site from owned or controlled sources, the greenhouse emissions associated with the energy generation are classified as direct scope 1 emissions. The same applies to companies, such as electricity utilities or suppliers, which control their energy generation facilities and sell all their power into the local grid. The greenhouse gas emissions from these electricity generation facilities are reported as Scope 1 emissions.

Put simply, scope 2 encompasses indirect emissions associated only with the generation of purchased or acquired energy.

Scope 3 Emissions

Scope 3 encompasses emissions that are not produced by the company itself and are not the result of activities from assets owned or controlled by them, but by those that it is indirectly responsible for, up and down its value chain. An example of this is when we buy, use, and dispose of products from suppliers. Scope 3 emissions include all sources not within the scope 1 and 2 boundaries.

Based on the financial transactions of a reporting company, the GHG Protocol divides the scope 3 emissions into upstream and downstream emissions:

- Upstream emissions encompass indirect greenhouse gas emissions within a company's value chain related to purchased or acquired goods (tangible products) and services (intangible products) and generated from 'cradle-to-gate' (the carbon impact of a product from the moment it's produced to the moment it enters a store; I'll explain more on this later).
- Downstream emissions include indirect greenhouse emissions within a company's value chain related to sold goods and services and emitted after they leave the company's ownership or control.

Bringing it to life

Scopes 1, 2, and 3 are somewhat abstract concepts and understanding how they get applied can sometimes be challenging for people to grasp. I'll try to bring it to life in a more familiar setting. Take a look at the picture below.

In this more familiar scenario, a weekend barbecuing session, the scope definitions would apply as follows:

In dark purple, Scope 1 emissions are created when the gas barbecue is fired up and uses the gas in the container attached. The barbecue owner has selected a rather large barbecue, which is possibly not as efficient as it could be depending on how much is being cooked and how often. These are direct choices that affect the quantity of emissions produced.

In light purple, Scope 2 emissions were created when the fuel supplier extracted the gas and processed it into the container now attached to the barbecue. The emissions were created indirectly, and you have purchased this fuel and will now use it to cook.

And finally, marked in yellow, are scope 3 emissions. These are related to emissions created by producing the physical barbecue itself. You didn't build the barbecue; you bought it from the local homeware store. These emissions were created during the production of the barbecue by the manufacturer and its subsequent storage and distribution to you. In this case, these are considered upstream scope 3 emissions as they relate to a purchased good.

Next time you have a barbecue, you might consider how all these elements come into play.

Emission factors

The release of GHGs into the atmosphere depends mainly on the activity and the product. In order to estimate GHG emissions per unit of available activity, we need to use something called an Emission Factor (EF); see the diagram on the next page[64].

[64] Source: Clim'Foot, climfoot_project.eu

An emission factor is a coefficient that allows you to convert activity data into GHG emissions. It is the average emission rate of a given source, relative to units of activity or process.

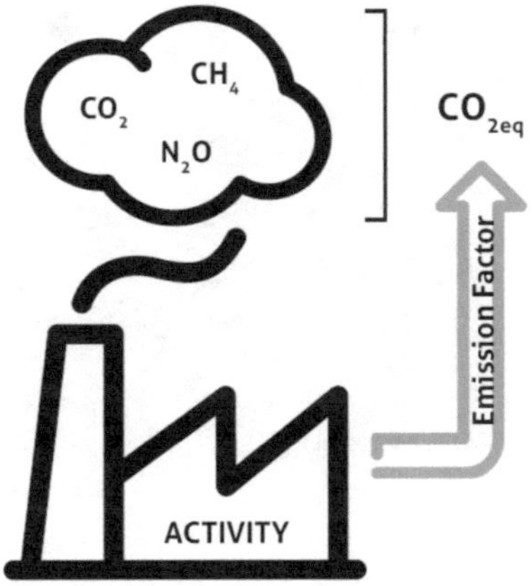

Emission = Factor x Activity

In general, emission factors are derived from measurements conducted on a number of sources that are representative of a specific emission sector. Human activities emit various greenhouse gases (GHG), and each has its own Global Warming Potential (GWP), a physical property of greenhouse gases as explained previously, representing their impact on the greenhouse effect. Each form of GHG emission is converted to its carbon dioxide equivalent, and as a result, the emissions of various gases can be compared. This all comes together to provide a consistent basis for measuring the emissions impact relative to the activity being undertaken.

Sustainable Development Goals (SDGs)

At some point, you might be asked, "Which SDG does your sustainable IT activity relate to?" And if so, answer: "Goal 13: Climate Action". But what does this mean, and why is it important?

The Sustainable Development Goals[65] of the United Nations (UN) are a blueprint for peace and prosperity for people and the planet, now and into the future. The Sustainable Development Goals (SDGs), which the UN adopted in 2015, are an urgent call to action for all nations, both developed and developing, as part of a global partnership. It is a recognition that ending poverty and other inequalities must go hand in hand with strategies that improve health and education and stimulate economic growth, all while addressing climate change and preserving our oceans, forests, and natural environments.

Science Based Targets

Science-based targets[66] provide a clearly defined pathway for companies to reduce greenhouse gas (GHG) emissions, thereby preventing the worst

---

[65] For more information go to the United Nations Development Programme website at undp.org

[66] For more information go to sciencebasedtargets.org

effects of climate change and securing business growth for the foreseeable future. Targets are considered 'science-based' if they are in line with what the most recent climate science deems necessary to achieve the goals set by the Paris Agreement. i.e., limiting global warming to well below 2°C above pre-industrial levels and pursuing efforts to limit warming to 1.5°C.

Science-based targets are based on the concept of a global carbon budget, which we discussed earlier. By accounting for the greenhouse gas emissions that have been emitted into the atmosphere since the beginning of the industrial revolution and having a good understanding of how these affect the climate, it is possible to estimate the level of additional emissions that can still be produced while still having a good chance of keeping global warming below 2°C.

The idea is that once you have a firm understanding of the concept of a global carbon budget and the various economic, social, and natural sources of greenhouse gas emissions, you will be able to plot the necessary reduction pathways in a fair and transparent manner. When implemented at the organisational level, this is considered a science-based target.

In essence, science-based targets rely on the expertise used to develop international scientific consensus and combine it with a transparent method for allocating a company's fair share of reductions. This means that a business can credibly demonstrate that it is contributing its equitable share to achieving the ambitions of the Paris Agreement.

There have been a wide range of concepts covered in the last two sections; don't worry if you are feeling slightly overwhelmed. I will revisit and expand on many of the terms later in the book to bring them to life in the context of forming a sustainable IT strategy. Next, I'll turn to technology and the key areas needed to decarbonise digital.

Knowledge Block Three: The ICT Sector

Information & Communication Technology (ICT)

Information and Communication Technology (ICT) is a term covering a diverse set of tools and resources used to transmit, store, create, share, or exchange information. These technological tools and resources consist of computers, the Internet (websites, blogs, and emails), live broadcasting technologies (radio, television, and webcasting), recorded broadcasting technologies (podcasting, audio and video players, and storage devices), and telecommunications (fixed or mobile, satellite, visual/video conferencing, etc.). The diagram highlights the main categories of ICT[67].

The OECD (Organisation for Economic Co-operation and Development) defines the ICT sector as "a combination of manufacturing and service industries whose products electronically capture, transmit, and display data and information[68].

[67] Source: TechTarget, techtarget.com

[68] OECD, 'Measuring the Information Economy', 2002

"The ICT sector is estimated to be worth approximately $550 billion annually"

Statista[69].

The GHG Protocol has recognised the significance and impact of the ICT sector on carbon emissions and developed the 'ICT Sector Guidance'. This provides a comprehensive set of guidelines and accounting methodologies for calculating the GHG (greenhouse gas) emissions of ICT products and services[70]. I'll cover more on the GHG Protocol ICT Sector Guidance in a later chapter, as it has become such a focal point in my life over the past two years.

Sustainable software engineering

Sustainable software engineering (or green coding) is a relatively new term that describes the techniques used by software programmers and developers to create code that minimises the energy consumption of software, thereby reducing its potential environmental impact. Instead of implementing new principles against existing practices, software engineers can now start by designing and developing code that strikes a balance between functionality and energy consumption.

Sustainable software engineering practices promote lean coding principles, in which minimal processing is required to produce the same result. A good example of this is when website developers prioritise smaller files and media of lower quality to achieve their goals.

In recent years, the acceleration of digital transformation has increased our reliance on data centres, and this will continue to grow exponentially as every

---

69 Statista, 'Global ICT market share 2013-2023', July 2023

70 GeSI, 'ICT Sector Guidance Built on the GHG Protocol Product Lifecycle Accounting and Reporting Standard' - go to gesi.org for more information

aspect of modern life becomes data-driven. Combine the growing demand for energy-intensive computing with a low-carbon and Net-Zero future, and it becomes clear why action across the ICT sector must take place.

The aim of sustainable software engineering is to reduce the quantity of energy required to power software operations through the technology value chain. Instructions for software applications and programmes are written by developers as lines of code, and as each line of code is processed, the device that processes this code uses energy that generates carbon emissions. The more code processed, the higher the emissions. Consider also that between 70 and 90 percent of software development[71] now employs non-tailored, open-source code containing numerous redundant sections of code written in the past, without the benefit of sustainable software engineering principles. This consumes even more processing power and generates unnecessary carbon emissions. The purpose of sustainable software engineering is to raise awareness of these inefficiencies and actively remove them to create energy-efficient software.

Software carbon intensity

Carbon Intensity (CI) is a measure of the environmental friendliness of a software component's production or operation. Typically, this will involve a conversion of the energy required to produce or operate the software component into carbon dioxide based on the emission factors relevant to the assessed software component.

A software carbon intensity calculation provides a true indication of an application's performance in terms of its environmental credentials. Once measured, it is possible to begin reducing and removing carbon emissions associated with the evaluated software component.

---

[71] Source: Dark Reading, R. Lemos, March 2022

Cloud computing

The marketing blurb tells you that cloud computing is the delivery of computing services over the Internet (or 'the cloud'), such as servers, storage, databases, networking, software, analytics, and intelligence, in order to provide faster innovation, flexible resources, and economies of scale. A key concept of cloud computing is that the location of a service and the hardware or operating system on which it runs are essentially irrelevant to the user. The reality from a sustainability perspective is that geography and location really matter, and I'll explain why in the remainder of this section.

The benefit of the cloud for businesses is that they can rent access to applications and storage from cloud service providers as opposed to owning their own computing infrastructure or data centres. Using cloud computing services enables businesses to avoid the upfront expense and complexity of owning and maintaining their own IT infrastructure and instead pay only for the resources they consume. However, consider that (1) this assumes cloud service providers (CSPs) can deliver this infrastructure more efficiently than you could yourself; (2) making the software that uses the cloud ready to fully benefit from cloud services requires considerable effort and expense; and (3) this is all now part of your scope 3 emissions, so you need to start understanding what is going on.

Logic does support the idea that by delivering similar services to a large number of customers, CSPs can realise significant economies of scale and efficiency. However, not fully understanding the way in which a CSP calculates the energy use and carbon impact of cloud service provision can result in consumers no longer having visibility and transparency over their consumption and, consequently, their energy-related emissions.

Is the cloud green?

Discussion on the environmental benefits and impacts of cloud computing dominates so much of the conversation that I have on decarbonising digital.

It is not surprising, as the big three cloud providers – Amazon Web Services (AWS), Microsoft Azure, and Google Cloud Platform (GCP) – provide two-thirds of the world's cloud computing. While all three have made commitments to decarbonise their data centres, none have stopped using fossil fuels. Due to the unavailability, inconsistency, or lack of transparency of credible emissions data relating to their operations, they frequently encounter difficulties in explaining their methods for reducing emissions. While CSPs may make the assertion that they utilise 'green energy', they actually rely significantly, if not exclusively, on offsets. This approach is heavily criticised by ESG experts as the use of the so-called 'market-based approach' conceals actual energy consumption. This method calculates emissions resulting from the electricity purchased by an organisation, which may be different from the electricity generated and consumed via the local power grid.

In order to provide a balanced view, let me answer the question from three different perspectives.

A positive perspective

Although the cloud does not have a perfect sustainability record, there is a lot of data available that indicates that cloud-based workloads are more eco-friendly than those operating in conventional data centres. And that trend is likely to become more pronounced moving forward as cloud providers invest further in renewable energy sources and as more workloads shift to environmentally friendly cloud-native formats.

In general, cloud-native technologies such as containers and Kubernetes consume less energy than conventional technologies such as virtual machines. But not all workloads operating in the public cloud are cloud-native, and cloud-native workloads can be run outside of the cloud (the term 'cloud-native' is misleading in this regard). Still, the cloud is the most logical location to host cloud-native workloads, and the widespread adoption of energy-efficient cloud-native technology has the potential to make the cloud even more sustainable.

Four arguments supporting the idea of a clean cloud

1. To prevent significant losses during the transmission of electrical energy over long distances, public cloud data centres are typically located close to the facilities that provide their power. Traditional data centres typically have far fewer location options. Due to superior hardware configuration, cloud computing data centres use less power to provide backup power and ventilation for their data centres. These data centres are designed on a large scale and constructed with energy efficiency, optimal utilisation, and temperature in mind.

2. In the past, businesses have used their own private data centres, which can result in low usage rates due to the purchase and installation of equipment in anticipation of server utilisation spikes. The cloud consolidates machine use by operating servers at high utilisation rates, thereby increasing productivity. When hardware remains idle, it is efficient. In contrast, due to highly utilised infrastructure, public cloud servers are quoted as being two to four times more efficient than traditional data centres.

3. Due to the high cost and time involved in upgrading servers, traditional data centre hardware is typically kept for a long time before being upgraded or replaced. Since the hardware in a public cloud tends to have much higher utilisation rates than traditional servers, it will most likely have a shorter lifecycle, necessitating a faster refresh time. It is also more cost-effective for public cloud servers to upgrade on a regular basis due to the improved energy efficiency that comes with new technology. The greater the hardware's energy efficiency, the more money the public cloud provider will save, resulting in less energy consumption over time.

4. Traditional infrastructure is difficult to maintain because it requires uninterruptible power supplies, refrigeration, and a lot of electricity. Moving basic software programmes to the cloud can result in significant energy savings. According to some cloud-supporting case studies, migrating business software to the cloud could reduce the energy consumption of these applications by as much as 87 percent[72].

A negative perspective

Where does the energy used to power your servers and data centres come from? Cloud giants may claim to be powered by 'green energy', but in reality, this is often achieved through carbon offsets. ESG bodies are already highly critical of cloud service providers who conceal their actual energy consumption by reporting Scope 2 emissions using the 'market-based approach.' This method estimates emissions from a company's purchased electricity, which may be different from the electricity generated and consumed by the local energy utility.

Claims of carbon neutrality and Net-Zero emissions based on offsetting are allowed to happen when a utility provider purchases off-site renewable electricity contracts or financial instruments such as Renewable Energy Certificates (RECs), which are tradable, non-tangible energy commodities. This distorts the true carbon intensity of electricity use, as each Renewable Energy Certificate verifies ownership of one megawatt-hour of electricity generated from a renewable source but is usually not supplied to the location where the consumption of that energy actually takes place.

Using RECS within a market-based approach can result in claims of zero carbon emissions for a cloud service provider, but this will not be the case. Using RECs to offset carbon emissions on paper to achieve 'Net-Zero' targets while retaining fossil fuel energy sources is a form of corporate

---

[72] Source: ISG 'How Cloud Computing Reduces Carbon Emissions' - Not validated by the author.

greenwashing and does not reduce emissions meaningfully. Only data centres that rely on location based renewable energy with no fossil fuels are moving towards true Net-Zero emissions.

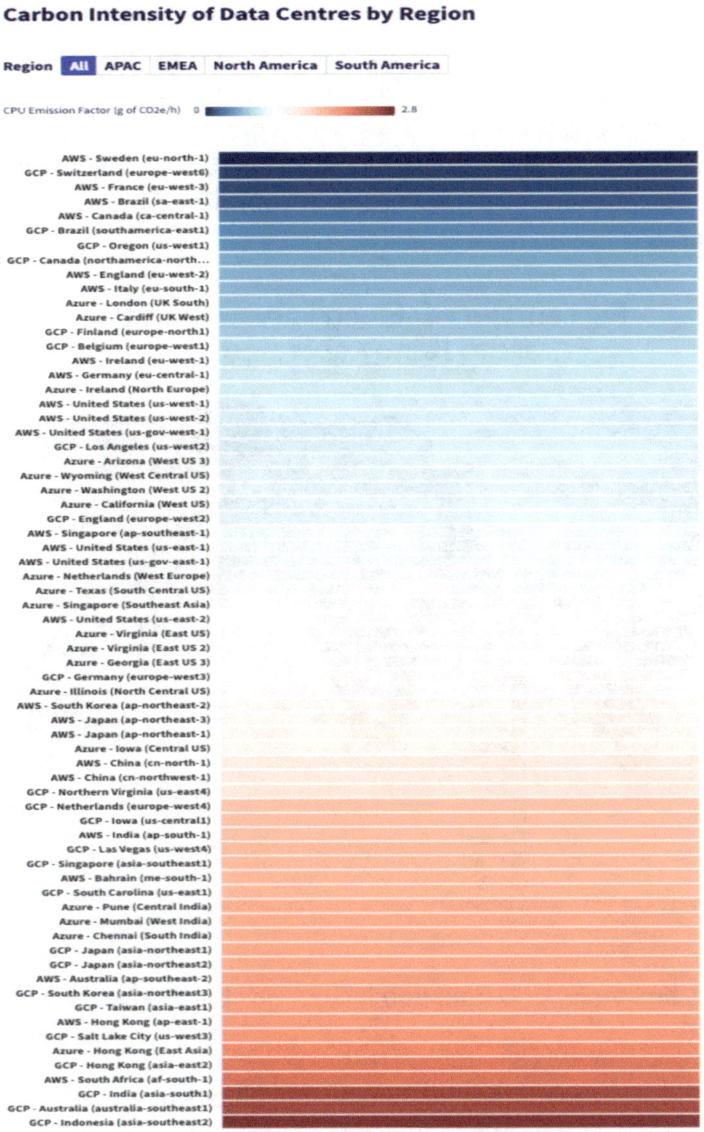

The chart[73] above shows the variation in carbon intensity between different sample data centres for each of the main cloud service providers in various geographic locations. It is clear that location and the supply of power from the local grid are major factors in determining the true level of sustainability.

Compare the previous chart with the map[74] below, and it is not surprising that the carbon intensity of data centres correlates closely with the geographic distribution of renewable energy generation. Consider this when determining the optimal location for your compute. And where procuring renewables is not possible due to local infrastructure limitations, consider ways to reduce the energy consumption of your digital products and services to compensate.

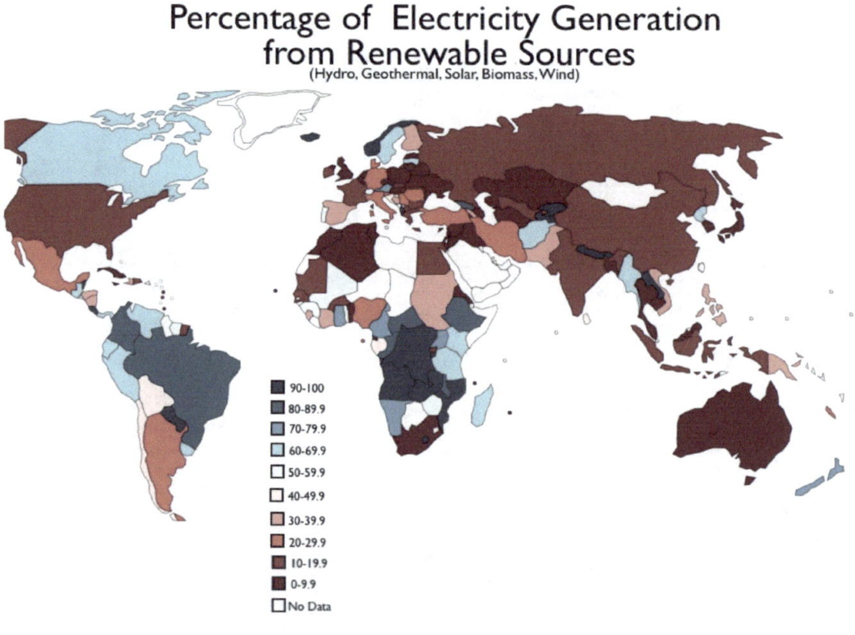

---

73 Source: Climatiq.io, CPU Emission factors via cloudcarbonfootprint.org and OEFDB

74 Source: GeoCurrents, Mapping Renewable Electricity Generation', Martin W. Lewis, using data from Wikipedia, geocurrents.info (undated)

A balanced view

Today, nearly all hyperscalers boast that their clouds are environmentally friendly and are advancing towards zero emissions. However, when it comes to the actual sustainability credentials of specific cloud deployments for specific businesses, it is not so clear-cut.

Cloud computing is characterised by deployment differences using a different mix of technologies, and when calculating the total carbon footprint, it is based on a complex array of factors. Public clouds can be a greener option when compared with more traditional approaches to computing, but the cloud may not be green for you, depending on the way you are using it.

For a balanced view, it is important to consider the following:

- In some instances, a poorly designed cloud deployment running on a very energy-efficient public cloud provider can generate more carbon than a fully optimised architecture running within a traditional data centre.
- The points of presence are where the cloud users are located. A cloud provider may point to a data centre linked to a wind or solar farm, but your cloud services, applications, and data may not be using that data centre. The data centre your company uses may be in an area, state, or country that uses a coal-fired power facility as its source of energy.
- Develop applications with efficiency in mind. Consider DevOps continuous testing services that determine if your application meets standards for security, performance, and stability, introduce a standard for power consumption, and incorporate it into your software development strategy.
- Developer priorities. Most developers don't yet focus on sustainability. Considering that they have to prioritise performance, security, stability, and functionality, asking them to code and test for power efficiency in support of sustainability is often seen as being too much effort. Make

sure they understand that it may only require a small amount of additional effort for a significant power optimisation gain.

- Applications and data storage systems can be 'power optimised' to substantially reduce cloud usage costs. Optimised workloads require less cloud resource allocation, run fewer power-hungry physical servers, and will cost you less as you will be using fewer resources. Ultimately, businesses can save additional money by delaying the purchase of new hardware if power-optimised application development is designed to use fewer resources.
- Challenge the hardware refresh cycles. Most CSPs are now stating refresh cycles ranging from 3 to 6 years, not the lower refresh rates previously claimed as a result of maximising utilisation. This indicates a lower level of efficiency than originally claimed and a realisation that software applications need to be designed to make the most efficient use of the cloud.

Summary

It is essential to evaluate and query the information you are given. Some hyperscale data centres unquestionably benefit from economies of scale, newer and more efficient infrastructural components, direct access to renewable energy, and advantageous geographic locations. Others simply do not. The important lesson to takeaway from this is to ask questions and be clear on the relative performance of the cloud service provider you are working with. Also, consider how you can optimise the energy efficiency of your software design and deployment at the code level to use less compute and energy in the first place.

A Recipe for Change

"There are 51 billion tons of greenhouse gasses emitted annually, what is the percentage reduction a proposed scheme offers? Technologies that will never exceed one percent should not compete for the limited resources we have for getting to zero,"

Bill Gates, 'How to avoid a Climate disaster', Sept. 2023

What follows is my recipe for decarbonising digital and creating a 'global digital reset', as well as my view on the size of the opportunity in front of us. Let's start by establishing a context for action with a calculation in line with Bill Gates' theory above. Let's assume that his approach to investment prioritisation is sound and that we can use the available data to determine a size of prize that we can start to seriously consider:

Total global emissions each year: 51 billion metric tCO2e per year[75]

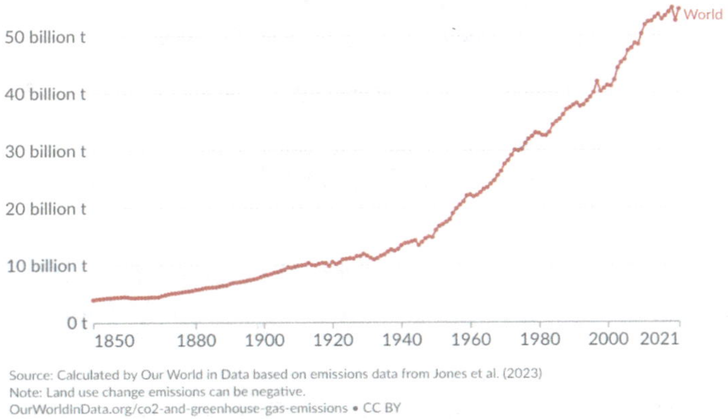

---

[75] Source: Our World in Data, after Jones et al, 2023

ICT is responsible for 5 percent of global emissions[76] = 2.55 billion metric tCO2e per year

The 5 percent figure for ICT is from a hardware-focused perspective and excludes emissions associated with the development and maintenance of software. The diagram below shows the distribution of ICT-related emissions and breaks this down in terms of hardware components and services[77]. It is the result of a research project undertaken in 2018 and remains one of the most thorough on this topic to date. Over the course of the next few pages, I'll explain my 'framework for measurement' and why the inclusion of software is so important to building a more complete picture of the carbon impact of ICT.

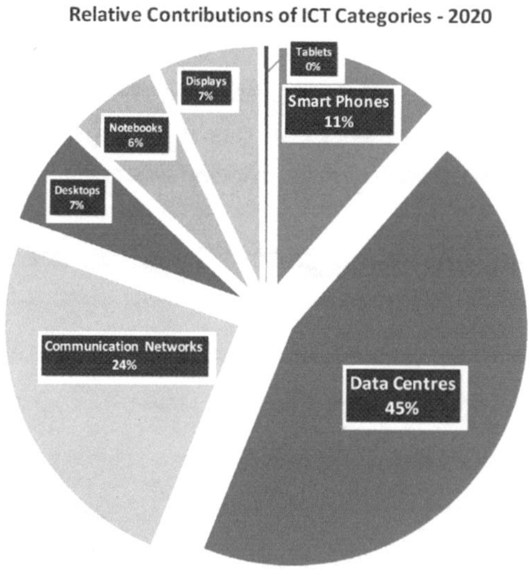

Before that, let's pause and ask ourselves whether we think it is possible to generate annual reduction opportunities of approximately 500 million metric

---

[76] UK Parliament, 'Energy Consumption of ICT', Sept. 2022

[77] Source: 'Assessing ICT global emissions footprint: Trends to 2040 & recommendations', Lotfi Belkhir and Ahmed Elmeligi, Jan. 2018

tonnes of CO2e, the 1 percent that Bill Gates suggests makes a decarbonisation lever worth investing in and pulling.

We have measured and observed opportunities to reduce the carbon impact of digital products and services and their technology value chains by an average of 56 percent at GoCodeGreen. Let's put this 56 percent, which sounds like a big number, into perspective. The United Nations[78] state that 'global carbon dioxide emissions need to be reduced by 45 percent by 2030 from 2010 levels, and reach Net-Zero emissions by 2050'. So the good news is that the scale of carbon reduction opportunity we are seeing is in line with achieving agreed global targets.

Let's put this 56 percent, which sounds like a big number, into perspective. The United Nations state that 'global carbon dioxide emissions need to be reduced by 45 percent by 2030 from 2010 levels, and reach Net-Zero emissions by 2050'. The scale of carbon reduction opportunity we are seeing is in line with achieving agreed global targets.

A framework for measurement

Why is including software in the scope of ICT's carbon impact important? My philosophy is based on the premise that software drives the demand and consumption of hardware, so if we can make software more energy efficient, then we can affect the efficiency of the entire technology value chain that supports it. The earlier ICT diagram gets recut (see next page) to show software at the centre of the pie and the opportunity to influence reduction in every category of ICT, whether that be through reduction in direct energy consumption, extending the lifespan of embodied emissions, or making more efficient use of hardware and devices.

---

[78] Source: UN SDG Goals - UN Statistics Division

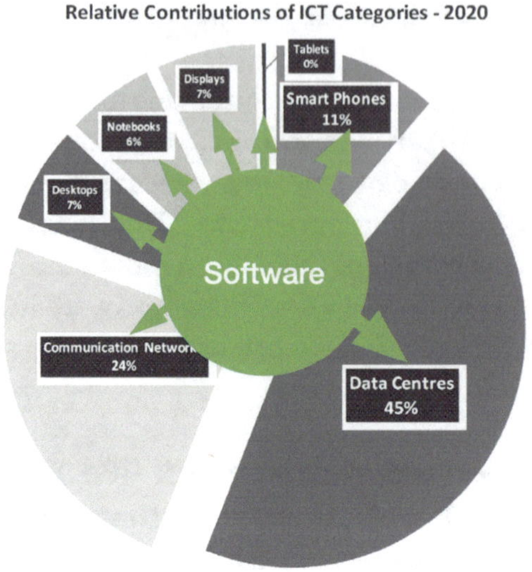

A vital component of my framework for measurement is the need to run consistent calculations all the way through the ICT categories and be able to set a solid foundation and context for taking action. I have seen too many organisations jump straight into low-level monitoring or one-dimensional calculations on single aspects of the ICT landscape without taking the time to consider the complexity and interconnectedness of their digital landscape.

Tackling this problem has been at the core of my work at GoCodeGreen, designing and creating measurement solutions to provide the first, really holistic view of an organisation's digital and ICT impact based on the GHG Protocol guidance and standards. This then allows for software product lifecycle stage assessments to help set overall decarbonisation targets all the way through an organisation's digital capabilities. By undertaking this context set, the organisation can then prioritise activities and support the use of the correct tools, methodologies, practices, and techniques to accelerate action.

The graphic below gives a summary of my 'Digital Decarbonisation Framework', and when this is aligned to my recipe for change and supported by the GoCodeGreen measurement solutions, I hope that you will start to see a pathway for turning the idea of decarbonising digital into a reality.

Digital Decarbonisation Framework

Set a context for success

1. Set the context to decarbonise ICT, through a **consistently calculated view** of the carbon impact across an IT estate

2. Screening to enable a **consistent and common understanding of total carbon impact across the 3 major ICT buckets** - desktop, network, datacenter & cloud – according to the GHG Protocol ICT Sector Guidance, Product and Scope 3 standards

3. The right **context to identify, prioritise and invest in the use of other tools** to support lower level action

4. Providing a **context for assessing / measuring and identifying action at software product level across lifecycle stages**

Set targets

Baselined context and applications

Actions identified and connected from software through to hardware

Objective key results (OKRs) and Key performance indicators (KPIs) set and tracked

Investment and cost reduction targets set and tracked

Take action

Hardware Action - Refresh to cleaner IT; Improve utilisation of existing IT; Rationalise amount of hardware being used; Migrate to cloud; Adopt Circular ICT approaches

\+

Software Action - Sustainable Software Engineering; Green Design; Software Energy Efficiency, People, Operational Activities; Tooling Choices; DevSustainOps; Software Supply Chain Decisioning

=

A DEEP AND SUSTAINED DECARBONISATION OF DIGITAL

A recipe for change

Let's convert everything we've discussed in this section so far into a recipe for change and bring it to life by estimating the range of possible target reductions (based on real client assessments completed by GoCodeGreen). There are three core ingredients that form the basis of my recipe:

1. Make software utilise existing embodied carbon more efficiently.

By doing so, we would seek to achieve the following:

- A 15 percent reduction through extending the life of embodied emissions in all devices by an extra year and therefore reducing demand for new device production.

- A 10 percent manufacturing-related reductions through supporting the use of circular ICT focused on recycling and reuse with software designed to operate efficiently on older or recycled devices.

2. Make software more energy efficient and reduce its energy demand.

By doing so, we would seek to achieve the following:

- A 45 percent reduction from more energy-efficient use of existing infrastructure components and cloud services using sustainable software engineering techniques, green design, improved utilisation of existing resources, and modernised cloud native applications and architectures.

3. Make the software itself more efficiently.

By doing so, we would seek to achieve the following:

- A 10 percent reduction gained from more efficiently building and maintaining digital products through embedding sustainability into DevOps (a methodology and set of practices and tools, to integrate and automate the work of software development and IT operations as a means for improving and shortening the systems development life cycle) and FinOps (a management practice promoting shared responsibility for an organisation's cloud computing infrastructure and costs) maturity pathways (a.k.a. 'DevSustainOps' which I'll cover later) and increased agility of delivery.
- A 10 percent reduction through using technology to reduce travel and increase collaboration.
- A 10 percent reduction using modern technology, tools and methods to engineer more efficiently (for example, NoCode, reuse, AI-based coding, new crowd-sourced and on-demand work practices, and more efficient automation, testing, and software development toolsets).

Following this recipe for change to breakdown the 56 percent carbon reduction that GoCodeGreen is observing, we might ambitiously calculate the following impact:

A 56 percent decrease from 2.55 billion tCO2e equals 1.43 billion metric tonnes of CO2e.

The table below summarises how everything might fit together in this bold reduction plan. Aligning the estimated reductions from the three ingredients into the GHG Protocol ICT Sector Guidance categories creates a bold view of the future for ICT. One that compares favourably with Bill Gates' 'Is it worth doing challenge' and the UN's 45 percent reduction by 2030. When mapped against the action reduction levers, categorised as 'make less new stuff', 'use what we have more efficiently', and 'make the software itself more efficiently', we begin to see the reduction opportunity ranges that could serve as the starting point for global ICT reduction and efficiency targets.

	Desktop Managed Services	**Communication Networks**	**Data Centres and Cloud**	**Totals**
Share of Total ICT related emissions	**31%** (790 million tCO2e)	**24%** (612 million tCO2e)	**45%** (1.148 billion tCO2e)	**2.55 billion tCO2e**
Bill Gates 'worth doing' challenge		**1%** (510 million tCO2e)		**510 million tCO2e**
Total Digital emission reduction opportunity per GoCodeGreen		**56%** (1.43 billion tCO2e)		**2.8x larger opportunity**
Making Software Utilise Existing Embodied Carbon more Efficiently i.e. make less new stuff		**25% reduction** 15% reduction through embodied emissions of devices extended by 1 year + 10% manufacturing related reduction through circular ICT		**127 to 357 million tCO2e**
Making Software more Energy Efficient and Reduce its Energy Demand i.e. use what we have more efficiently		**45% reduction** Through software that makes more efficient use of existing network and data centre infrastructure		**230 to 647 million tCO2e**
Making Software more Efficiently i.e. make the software itself more efficiently		**30% reduction** 10% by deploying processes that drive agility and velocity, 10% through automation and use of new technologies and development tools that speed delivery times and 10% though reducing the need for travel whilst improving collaboration		**153 to 428 milion tCO2e**

Of course, these calculations are intentionally oversimplified; their purpose is to help you imagine the magnitude of the challenge and opportunity. In this context, Gates' 1 percent reduction, the equivalent of 500 million tCO2e (the same carbon emissions as Brazil[79]), appears to be well within reach. The fact that we are even talking about placing a value and an estimate on this scale of impact and opportunity for digital in real numbers is a step forward.

Even more encouraging is the fact that, at the micro-level, the target reductions are not merely estimates. GoCodeGreen has observed this scope of opportunity in real-world situations, and all of these reduction levers are currently available to us. What we are talking about here is not one of those 'technologies that will never exceed one percent', and we do not require a miraculous new technology that will take a decade to test, prove, manufacture, and implement. This is a NOW opportunity.

But let's also be realistic about timing; taking anything like this scale of action will need to be phased, the reductions can't be delivered all at the same time, and there will be advancements in physical hardware and the development of the next generation of ultra-efficient devices that will need to flow through the supply chain to enable this next order of scaled reduction opportunity. In subsequent phases of decarbonisation, increased renewable energy supply and new, more efficient ICT manufacturing processes and distribution will result in greater carbon reductions. But there is no reason to delay beginning real action that can have an immediate impact and set the groundwork for even greater results when those new production methods become available.

Regardless of the final target number, what we do know is that it needs to be big and directionally scaled towards the hundreds of millions of tonnes of CO_2e in order to achieve the scale of impact needed. In order to support this scale of reduction, we will need an adjustment to the current business and commercial models of technology companies. And by the way, this doesn't necessarily mean lowering profitability.

---

[79] Source: ourworldindata.org, referencing country level carbon emissions from 'Global Carbon Budget 2022'

Bill Gates talks about an approach centred around the concept of a 'Green Premium,' which is based on the additional cost of a product or service that is produced in an environmentally responsible and sustainable manner. Gates argues that immediate adoption is required if the green premium is minimal or even negative. If it is high, commercially problematic, or creates short-term competitive challenges, then government intervention may be required to enforce or incentivise its use[80].

For ICT, green premium pricing could be based on the cost of targeted provision of compute so that we only use what we need, paying higher prices for circular or high energy efficiency devices, or software companies recovering the cost of decarbonising and maintaining the lower energy performance of digital products via increases in licensing and maintenance fees. All of these increases would be balanced by a decrease in energy consumption from achieving greater efficiency.

I believe that the decarbonisation of digital is a low-green premium and one that requires immediate adoption. The positive news is that a market is already forming, and we are starting to observe a shift in consumers' understanding, acceptance, and desire to pay for green premium pricing.

This opportunity's size relative to its cost and potential energy savings is significant and cannot be ignored.

Over the course of the remainder of the book, I will further examine the ingredients required to make this framework and recipe for change a success. This will include enablers such as green software, green design, and sustainable IT leadership, as well as the critical role that people and culture play in binding these ingredients together. These elements combined will help us achieve a much-needed global digital reset.

---

[80] The Register: 'Bill Gates on climate change: Planting trees is not the answer, emissions need to be zeroed out to avoid disaster', Feb. 2021

Part Two: Green Software

Sustainable Software Engineering

Sustainable software engineering (also known as sustainable software development or green coding) is gaining popularity daily. This approach seeks to design software that improves an application's efficiency and reduces its energy and consumption demands. These 'green' software products would collectively require less compute and data storage and lower the long-term global impact of digital.

Sustainable software engineering is not only better for the environment but also aids businesses in creating more efficient products and services. The demand for carbon-neutral and energy-efficient digital products is growing rapidly, and this is becoming an important trend driving the modernisation of existing software solutions. In response, many software companies have started to develop more energy-efficient digital services, and it is becoming acknowledged that energy consumption will become one of the primary metrics considered when consumers select a new digital product or service. There is also mounting pressure from customers resulting from new reporting standards to better understand the scope 3 impact of their purchased digital services.

The impact of energy-efficient software for a business

One trend I have observed is that businesses are investing in software with higher sustainability standards and are starting to consider the environmental credentials of both the software supplier and the products they produce. Small and large businesses alike are searching for software that is simple to deploy, consumes less energy, occupies less space in the cloud, and provides more functionality. They are increasingly supporting companies that are investing in software development that can demonstrate they are transitioning towards a better, more sustainable digital world.

The information provided in Part One makes it clear that we are living in a digital age, and this is contributing to the damage we are causing to the environment. When it comes to software, this may not always be immediately apparent, but we can make choices to significantly reduce this impact. Part Two will bring some of these choices to life through a set of sustainable software engineering principles and techniques that will help decarbonise digital products and services. The next few sections are intended to be technical, as I need to properly explain the types of actions required to make software more energy efficient. If you are not a software engineer or don't have a coding background, then try to persevere with the content, as the concepts are core to how we decarbonise digital.

The rise of the sustainable software engineer

If we employ the same software development processes as our industry did 20 or more years ago, when coding was restricted to strict lengths and sizes and constrained by limited compute power, coding can generally become greener almost immediately.

Over the years, our ability to write longer and less refined code, with an increasing number of shared frameworks and libraries, has resulted in almost limitless programming capabilities. This is in part due to the pressure to create solutions in ever-decreasing cycle times. 'Bloatware' has resulted from the combination of low-cost, immensely abundant computing resources and the resulting proliferation of software.

While this more relaxed approach to coding has undoubtedly contributed to the pace of digital transformation and the modernisation of organisations and industries, we have reached a critical juncture in which we must now reevaluate our methods.

Achieving a Net-Zero future requires a balance between programming quality and energy consumption. For example, text, image, and video file sizes can almost always be reduced. A simple media compression utility can reduce file sizes by more than fifty percent, thereby enhancing the user experience, accelerating navigation, and conserving energy. Website developers need to know when to prioritise high-quality multimedia and when smaller files can achieve the same goals. Having fewer large files reduces loading periods, which helps businesses with both search engine optimisation (SEO) and their sustainability initiatives.

If you are involved in building digital products and services, it is essential to understand what can be done to support the transition to sustainable software engineering practices and techniques.

Now is the time for the sustainable software engineer, and the architects of the next generation of energy- and carbon-efficient digital solutions.

Practical - Sustainable Software Engineering Self Assessment

If you are a software engineer or technically savvy, complete the Sustainable Software Engineering Self Assessment. This will provide you with a rating based on your understanding and execution of the principles and techniques described in this part of the book. The ratings are calculated using insights from GoCodeGreen's assessments across multiple industry sectors and organisations, as well as the engineering weightings and efficiency factors used in our more comprehensive product assessments.

You can access the self-assessment using the QR code above, or alternatively, visit: www.gocode.green/sustainabilityassessmenttools

The Principles of Sustainable Software Engineering

The five principles of sustainable software engineering outlined below were developed to be a concise set of guidelines that, when followed, can help reduce energy consumption and carbon impact. The objective was to make them easy to use on a regular basis and to provide a foundation for the more detailed sustainable software engineering techniques discussed in a later section.

The five principles cover:

Principle 1: Engineering Energy Efficiency
Principle 2: Carbon Efficiency
Principle 3: Hardware Efficiency
Principle 4: Data Efficiency
Principle 5: Measurement

1. Engineering energy efficiency

To reduce the energy consumption of software and the technology value chain that supports it (e.g., infrastructure, hosting, devices, networks, etc.), it is essential to develop energy-efficient applications. All software execution consumes electricity. By making applications more energy-efficient, the Engineering Energy Efficiency principle aims to reduce software's electricity consumption and the resulting carbon emissions:

A sustainable application is accountable for the amount of energy it consumes and is designed to consume as little energy as possible.

2. Carbon efficiency

Carbon dioxide (CO_2) is the predominant Greenhouse Gas (GHG). It is essential that we reduce the quantity of greenhouse gases in the atmosphere immediately. If we do nothing, we will reach a critical point that will result in irreversible climate change. We need to stop adding carbon to the atmosphere and move towards achieving Net-Zero emissions. In a digital world, the software being developed can play a significant role in reducing the demand for energy and, consequently, the amount of carbon dioxide emitted into the atmosphere:

Carbon-efficient software minimises the amount of carbon emitted per unit of work and, whenever possible, consumes electricity with the lowest possible carbon intensity.

3. Hardware efficiency

There are two critical concepts when thinking about hardware efficiency principles: embodied carbon and energy proportionality.

Embodied carbon

In some instances, the embodied carbon cost of a device can be significantly higher than the carbon cost of the electricity used to power it. Therefore, it makes sense to attempt to increase the lifespan of a hardware device (thereby increasing the usefulness of the carbon embodied in the device) by creating software that is more hardware-efficient. Focusing on the development of applications that can operate on older hardware can play an important part in increasing the hardware's efficiency, as outdated hardware is frequently retired due to its inability to handle modern workloads:

The embodied carbon cost of a device can be greater than the cost of the electricity used to power it; therefore, it is essential to increase its lifecycle by developing software that is more hardware-efficient.

Energy proportionality

Energy proportionality is the ratio between the amount of energy consumed by a computer system and the rate at which useful work is performed. Computer hardware becomes more efficient at converting electricity into useful computing operations the more it is used; therefore, the most efficient and environmentally friendly strategy is to perform your workload on as few servers as possible with the highest rate of utilisation:

Developing applications that are compatible with obsolete hardware can also boost performance. Energy proportionality, the ratio of energy consumption to useful work, suggests that the most efficient strategy is to employ the fewest number of servers feasible.

4. Data efficiency

Reducing the quantity of data and the distance it has to travel can reduce energy demand and carbon emissions. A network consists of hardware components, including switches, routers, and servers. All of the network's processors and network equipment consume electricity and have embedded carbon:

When data is transmitted across the internet, it passes through numerous network devices, each of which consumes electricity. As a result, any data transmitted or received via the internet produces carbon emissions. Reduce the volume of data being transferred and make any required data transfers as efficient as possible.

5. Measurement is key

If it cannot be measured, it cannot be changed or optimised. It is essential to be able to measure the carbon emissions associated with software development and operation in order to understand the full impact of your digital product or service:

By choosing an appropriate measurement tool, you will be able to identify the key actions you can take to reduce the carbon cost of your software, as well as evaluate and validate those actions in a repeatable and consistent manner.

The Techniques of Sustainable Software Engineering

This section describes the twelve key sustainable software engineering techniques that underpin the previously discussed principles. These approaches should be added to your software development toolkit in order to support the development of greener, more energy-efficient software. Some can be implemented immediately and on a standalone basis, while others should ideally form part of a broader strategy for designing and building energy-efficient code.

The twelve techniques over:

Technique 1: Software Power Consumption
Technique 2: Data Caching Efficiency
Technique 3: Data Exchanges
Technique 4: Data Compression, Aggregation and Retention
Technique 5: Image Sizing
Technique 6: Unused Features
Technique 7: Logging, Looping and Polling
Technique 8: Power Adaptation
Technique 9: Computational Accuracy
Technique 10: Monitoring Energy Consumption
Technique 11: Programming Language Efficiency
Technique 12: Reuse

1. Software power consumption

The underlying cause of ICT energy consumption is the software that commands hardware to start processing. Software can be designed to direct hardware to execute its tasks with less energy consumption. Techniques can concentrate on: (1) mapping the energy consumption that is caused by the use of software and then searching for opportunities to reduce the software's energy usage; and (2) the development of tools to manage energy usage.

The optimal strategy is to write the software in such a way that the hardware uses less electricity in the first place. If the software is developed to allow it to run on various hardware configurations, the efficiencies are more likely to be replicated with each use, so identifying application features that have been calculated to be energy and resource intensive is a good place to start. Follow this by employing sustainable software engineering principles and techniques to reduce power consumption and maximise efficiency.

Power consumption measurement

Power consumption techniques are centred on integrating energy efficiency into software, preferably at the beginning of the energy-use chain. It is not enough to just instruct the hardware to enter sleep mode after the software has completed its execution (though sleep mode as a technique can be useful and is explained later in this chapter). A more effective approach is to write software in such a way that the hardware it runs on consumes as little energy as possible. This means that every instruction requiring energy is as efficient as possible. In the past few years, significant research has been conducted on measuring and tuning power consumption at the software level, with case studies demonstrating that this can result in energy savings ranging from 30 to 90 percent[81]. By making the effort to reduce energy use, it is possible to significantly reduce carbon emissions too.

---

81 Source: Network World, 'IT Best Practices', L. Musthaler, Dec. 2014

To understand the energy consumption of a software product or specific features or components within it, you need to measure the power consumption of the code. In practice, these techniques are costly to implement, but when undertaken, they can allow for comparisons of energy efficiency between various software applications, software architecture designs, or different source code. The practicalities can be challenging, requiring different test bench setups and repeated cycles of analysis, but taking this approach can help determine when scenarios are appropriate for further investment in terms of both time and money.

There are different methods available for measuring the energy consumption of your code, and there is no single universal approach since different platforms require different measurement strategies. Some tools, for instance, are only compatible with specific processor types, and others with a particular operating system. The two methods described below help provide a starting point for choosing the right approach based on your coded product's specific requirements.

1. Use a power monitor

Using a power monitor is the most accurate method for measuring energy consumption; these are hardware tools that connect to the power source of a device and measure the actual power being used at any given time. Despite their precision, power monitors are extremely difficult to install. Occasionally, they may even require you to make modifications to your computer hardware. It will also be necessary to time-synchronise the power data with the execution of your software, making the task even more challenging.

Some power monitors include a software library for controlling their work, for example, starting, stopping, and collecting readings, and the complexity increases if you are working with a team of developers who need to conduct multiple energy tests in parallel from different workstations. Using power monitors is a complex, code-specific, and expensive exercise, but it can provide detailed insight into your code's energy demands and, consequently, its carbon footprint.

2. Use an energy profiler

An alternative method is to use an energy profiler. These do not require any specialised hardware or power sensors, and they typically include a calculation model for estimating the power consumption of various hardware components. Based on which components are active during execution, the energy cost is estimated by the profiler. Energy profilers are typically distributed as simple-to-use utilities that can be executed on any device to calculate total energy consumption. Another benefit is that timing synchronisation with software execution is simplified.

If you choose to select an energy profiler, it is important to remember that they rely on estimation and only operate under very specific conditions. As a result, there is no one-size-fits-all energy profiler, and finding one that works for a particular use case can in itself be a challenging task.

When measuring energy consumption, it is important to not separate the software being evaluated from the system on which it is operating. To ensure that results are not compromised, it is crucial that the device's workload be reduced to the absolute minimum and that no unnecessary applications or processes are running, as these could distort your final results.

Measuring power consumption

Unlike energy consumption, which is measured in joules, power is measured in watts. It is essential to understand the difference in order to read and interpret the results from power monitors and energy profilers. To compute the energy consumption (E), in joules, you need to multiply the average power (P), in watts, by the elapsed time in a given sample (\Delta t) in seconds:

$$E = P.\Delta t$$

Over the remainder of this section, I'll describe several common energy measurement tools that can be used. Learning how these tools work will help you understand the underlying methodology and approach being used to determine the energy needs of your code. And when used to measure one or more processes carried out by your software, it is possible to begin to understand the energy associated with a feature or set of features in your digital product.

Powerstat

If you are developing software in a Linux environment, you should consider using Powerstat to measure energy consumption. Similar to other tools, Powerstat utilises an Intel library called RAPL, or Running Average Power Limit. It provides a straightforward interface for command-line execution. You can conduct a power measurement with the following command:

```
sudo powerstat -R 60 1
```

You can readily alter the default measurement duration of 100 seconds (10 samples at 10-second intervals) when calling the tool. In the given configuration example, 60 samples are taken at 1-second intervals.

The output is a table with multiple statistics per sample. The Watts column in the table shows how much power each sample used on average. At the end of the output, Powerstat displays summary results, which include the average CPU power in Watts, for example:

```
Summary:
CPU:  8.38 Watts on average with standard deviation 2.53
Note: power read from RAPL domains: package-0, uncore, package-0, dram, core, dram, psys.
These readings do not cover all the hardware in this device.
```

In this instance, the summary tells us that the average power consumption (notice the distinction; this is not energy consumption) of the execution was 8.38 W. The total energy consumption can then be calculated based on this electricity consumption. In this example, the measurement was conducted for sixty seconds, and the total energy demand is therefore 8.38W x 60s = 502.8J.

Intel Power Gadget and PowerLog

These tools are often used for Macs and Windows on Intel processors. To collect energy consumption data, click on the menu Logging > Log to File to start measuring and again to stop the measurement process. The power data will be collected and stored in a CSV file.

If you open the CSV file, you will find several columns. Some are straightforward, for example: System Time, Elapsed Time (sec), Processor Power_0(Watt), while others tend to be more complex, for example, RDTSC and GT Requested Frequency(MHz).

Scroll all the way down to the bottom of the file, and you will locate a summary table with the following key attributes:

- Total Elapsed Time (sec). The total time in seconds in which power data was being collected.
- Cumulative Package Energy_0 (Joules). The total energy consumption of the processor.
- Cumulative DRAM Energy_0 (Joules). The total energy consumption of the volatile memory.

In addition, you can also use Intel PowerLog, a command-line interface tool inside the Power Gadget bundle. Running a command to execute will also generate the CSV file 'results.csv' that is identically formatted to the Intel Power Gadget data above.

Executing this collection whenever a particular use case of your software runs will provide data to help support the optimisation of your code's energy efficiency. Note that the Power Gadget UI requires manual activation, so it is not the best option for assessing software that requires scaled, automated energy assessments.

Perf

Perf provides a method for determining a software program's energy consumption in a Linux environment. Perf is a command-line utility that also wraps Intel's Running Average Power Limit (RAPL). It enables the collection of energy measurements from the cores, GPU, pkg (core and uncore components), psys, and RAM of a computer system.

You can obtain an application's energy consumption with the following command:

```
sudo perf stat -e power/energy-cores/,power/energy-ram/,power/
energy-gpu/,power/energy-pkg/,power/energy-psys/ sleep 5
```

Perf will provide output that looks something like:

```
 Performance counter stats for 'system wide':

          3.59 Joules power/energy-cores/
          8.18 Joules power/energy-ram/
          1.63 Joules power/energy-gpu/
         12.74 Joules power/energy-pkg/
         51.10 Joules power/energy-psys/

      5.001965465 seconds time elapsed
```

In this example, the energy consumption of a Linux command with a 5-second slumber setting is being measured.

Perf is highly configurable and has broad coverage, with the ability to measure almost any executable in any programming language. Moreover, Perf offers the -r command-line argument that allows a user to specify how many times to measure an application.

```
sudo perf stat -r 5 -e power/energy-cores/ sleep 5
```

For example, the preceding command will implement the sleep 5 command five times and report the average energy consumption as well as the difference between the measurements as a percentage:

```
 Performance counter stats for 'system wide' (5 runs):

          4.44 Joules power/energy-cores/       ( +- 1.52% )

       5.002405 +- 0.000272 seconds time elapsed  ( +-  0.01% )
```

PowerTOP

PowerTOP is a tool for measuring and monitoring energy consumption on AMD-powered devices. Use cases for PowerTOP can go beyond straightforward energy consumption measurements as it has a broader range of features; for example, it provides an interactive mode that allows users to fine-tune their Linux system's power management settings.

A good start command for PowerTOP is:

```
sudo powertop --csv=output.csv -t 20
```

Where the --csv option prevents PowerTOP from entering the interactive mode and sets the output to be stored in the file 'output.csv', and the option -t 20 sets the measurement to run for 20 seconds.

Although the option --csv is set, the output is a full output of text mixed with tables formatted in a 'CSV-like' format. This includes a fine-grained estimation of the power consumption of each process in the system. As an example, see the sample output below:

```
* * *  Overview of Software Power Consumers  * * *

Usage;Wakeups/s;GPU ops/s;Disk IO/s;GFX Wakeups/
s;Category;Description;PW Estimate
172.9 us/s; 24.9;;;;Timer;tick_sched_timer; 98.5 mW
296.0 us/s; 22.9;;;;Process;[PID 313728] /usr/bin/containerd ; 90.8 mW
 96.5 us/s; 16.2;;;;Process;[PID 11] [rcu_sched]; 64.2 mW
225.5 us/s; 14.1;;;;Process;[PID 313736] /usr/bin/containerd ; 56.0 mW
409.9 us/s; 12.3;;;;Process;[PID 361909] /usr/bin/vmtoolsd ; 49.4 mW
232.0 us/s; 12.1;;;;Process;[PID 313737] /usr/bin/containerd ; 48.3 mW
 24.5 us/s;  4.9;;;;kWork;fb_flashcursor; 19.4 mW
 21.4 us/s;  4.8;;;;kWork;vmw_fence_work_func; 19.0 mW
156.0 us/s;  3.9;;;;kWork;psi_avgs_work; 15.8 mW
309.8 us/s;  3.4;;;;kWork;vmw_fb_dirty_flush; 13.9 mW
 29.4 us/s;  3.3;;;;kWork;gc_worker; 13.1 mW
311.6 us/s;  1.7;;;;Timer;hrtimer_wakeup; 7.19 mW
 23.8 us/s;  1.2;;;;Interrupt;[17] ioc0; 4.97 mW
 39.7 us/s;  1.0;;;;Process;[PID 811] /usr/sbin/ntpd -p /var/run/ntpd.pid -g -u
113:117 ; 4.20 mW
 15.5 us/s;  1.0;;;;Timer;watchdog_timer_fn; 3.97 mW
  4.4 us/s;  1.0;;;;kWork;vmballoon_work; 3.95 mW
  6.0 us/s;  0.9;;;;kWork;mpt_fault_reset_work; 3.76 mW
 15.8 us/s;  0.8;;;;kWork;vmstat_shepherd; 3.38 mW
  1.1 ms/s; 0.15;;;;Process;[PID 377852] powertop --csv=output.csv -t 20 ;
2.33 mW
[...]
```

Included within this rich data set are data points for energy consumption. Based on the previous example, search the output file for the following line:

The system baseline power is estimated at: 5.43 W;

This means that for this use case, on average, the processor and main memory used 5.43 W of power during its 20 seconds of execution. According to the earlier power consumption formula, the total energy consumption is calculated by multiplying the average power by the duration of the measurement as follows:

$$E=P.\Delta t=5.43W\times 20s=108.6J$$

PowerTOP is a useful energy profiling tool that covers a range of energy consumption techniques and is especially suited for measuring battery-powered devices.

Likwid

Utilising Likwid is another method for gathering energy and power measurements from Intel processors. Intel's RAPL interface is used by Likwid to retrieve energy and power measurements from a CPU.

By executing the following command, you can obtain energy and power measurements for the device:

```
likwid-powermeter sleep 5
```

The energy consumption output is provided in a format that will look something like this:

```
--------------------------------------------------------------------------------
CPU name:     Intel(R) Core(TM) i7-8565U CPU @ 1.80GHz
CPU type:     Intel Kabylake processor
CPU clock:    1.99 GHz
--------------------------------------------------------------------------------
--------------------------------------------------------------------------------
```

```
Runtime: 6.96028 s
Measure for socket 0 on CPU 0
Domain PKG:
Energy consumed: 24.9688 Joules
Power consumed: 3.58732 Watt
Domain PP0:
Energy consumed: 11.2714 Joules
Power consumed: 1.61939 Watt
Domain PP1:
Energy consumed: 2.91577 Joules
Power consumed: 0.418916 Watt
Domain DRAM:
Energy consumed: 11.8179 Joules
Power consumed: 1.6979 Watt
Domain PLATFORM:
Energy consumed: 77.1105 Joules
Power consumed: 11.0786 Watt
--------------------------------------------------------------------------------
```

In contrast to Perf, Likwid does not provide an option such as -r to repeat an energy consumption test multiple times. However, it does provide other options, such as temperature monitoring per thread, that might be useful based on the assessment being conducted (e.g., assessing whether overheating of the device is occurring during the run time of the application).

Nvidia-SMI

The nvidia-SMI utility provides access to data related to a GPU's power consumption as well as information such as memory usage and temperature of the device. This is especially helpful for measuring the power consumption of GPU-intensive applications like training a deep learning model, viewing a video, or playing a high-definition computer game.

After locating the suitable package manager, execute the following commands:

```
sudo apt update
sudo apt upgrade
apt install nvidia-[driver_number]
```

Then restart your system and execute the nvidia-SMI command in your terminal. Measure the GPU power consumption of an application as follows:

```
+-----------------------------------------------------------------------------+
| NVIDIA-SMI 470.57.02    Driver Version: 470.57.02    CUDA Version: 11.4     |
|-------------------------------+----------------------+----------------------+
| GPU  Name        Persistence-M| Bus-Id        Disp.A | Volatile Uncorr. ECC |
| Fan  Temp  Perf  Pwr:Usage/Cap|         Memory-Usage | GPU-Util  Compute M. |
|                               |                      |               MIG M. |
|===============================+======================+======================|
|   0  Quadro P400         On   | 00000000:01:00.0 Off |                  N/A |
| 34%   27C    P8    N/A /  N/A |      1MiB /  2000MiB |      0%      Default |
|                               |                      |                  N/A |
+-------------------------------+----------------------+----------------------+

+-----------------------------------------------------------------------------+
| Processes:                                                                  |
|  GPU   GI   CI        PID   Type   Process name                  GPU Memory |
|        ID   ID                                                   Usage      |
|=============================================================================|
|  No running processes found                                                 |
+-----------------------------------------------------------------------------+
```

At this point start the application, in this example it is training a deep learning model, and obtain query information on the power usage of the GPU.

```
python trainDeepLearningModel.py &
nvidia-smi --loop-ms=1000 --format=csv --query-gpu=power.draw
```

The preceding commands will initiate a deep learning model training cycle in the background and then print the GPU's power consumption every second (--loop-ms=1000) in the foreground. The --format=csv option allows for data collection in CSV format. Nvidia-SMI supports a range of use cases and contains a wealth of online documentation.

Energy consumption comparison techniques

Selecting an appropriate use case is the starting point for energy consumption measurement if you want good data. Taking a feature-based approach can help because they are readily identifiable units of work or activity, and as a result, a wider variety of stakeholders are more likely to feel engaged with the findings and outcomes. A feature-level analysis can also assist in prioritising optimisation efforts and placing energy efficiency at the top of the list. But measuring energy consumption at the feature level can be more challenging due to difficulties in defining its boundaries; therefore, measurement of methods, classes, and other software units is often a more straightforward approach.

If getting granular at the feature level proves too complex, consider comparing the energy consumption of two different versions of your software in the same use-case scenario as an alternative method. Follow these steps:

- Create a reproducible scenario of the execution of your software, e.g., using a unit test framework.
- Execute the scenario in a version of your software, using the energy profiler to measure the energy consumption.

- Identify code that you suspect has higher energy consumption and measure performance.
- Execute the same scenario with new versions of optimised code and compare the energy data between versions.

This analysis provides insight into how software energy efficiency is improving as you optimise at the code level. This type of analysis can be incorporated into your pipeline for continuous integration.

You can also compare the energy consumption of various use cases in a particular version of the software, for example:

- Creating reproducible scenarios for different use cases of the software.
- Execute and measure all the scenarios.
- Compare the energy consumption of each of them.

This analysis can help you determine which portions of the code require optimisation.

Summary

Utilising power monitors and energy profilers can help you measure the energy consumption of virtually any desktop or server software, but it is a resource-intensive exercise. You must also consider the energy and carbon footprint of these tests, as well as the likely trade-off between optimisation-related improvements and the expense associated with carrying out the measurement. Selecting the right use case and balancing when to complete code-level measurement is key. It is good practice to keep a watchful eye on energy efficiency, and understanding these techniques and tools will help you determine when detailed energy measurement is appropriate.

2. Data caching efficiency

The difference between memory speed and processor speed is referred to as the memory gap. This discrepancy led to the development of cache memories, which are small amounts of fast but expensive memory used to improve performance by holding required data closer to the processor than main memory. However, because cache memories are optimised for performance rather than energy consumption, cache memories can account for a significant amount of the total energy consumed by microprocessor-based architectures.

Adopting an efficient caching policy can eliminate any unnecessary energy demands placed on the network and hardware as a result of data calls from your software. Even temporary storage results in network load and, therefore, energy consumption.

Data caching

Caching performance and a strategy covering the overall use of processing and storage during code development are really important for energy efficiency. By way of illustration, examine the following commonly used PHP command (a server-side scripting language embedded in HTML that allows web developers to create dynamic content and interact with databases):

```
function getArticles() {
    // Check if cache has data
    if ( Cache::has('articles') ) {
        $articles = Cache::get('articles');
    } else {
        // Populate articles data from database
        $articles = Articles::all();

        // Put articles data to cache and set TTL to 60s
        Cache::put('articles', $articles, 60);
    }
    return $articles;
}
```

cache2.php hosted with ❤ by GitHub view raw

This example uses a 'cache aside' strategy. It is used in a variety of applications, particularly those with a greater proportion of reading requests than writing requests. As it is a straightforward caching mechanism, many developers copy and clone the code. But with power efficiency in mind, a key question to ask is, 'Do I really need to cache this or all of the data?' This is particularly important when the presence of multiple data types complicates an application. Not all HTML pages, post-calculation data, database-retrieved data, and shared configuration data need to be cached.

To maximise efficiency, it is crucial to consider the following factors when determining if the cache is necessary:

- Time or resource consumption: time, CPU usage, disk IOPS, network bandwidth, file descriptor, etc. required to perform a task. Using caching is expected to reduce time or resources if the task needs to be re-executed.
- The result can be reused multiple times; the cached output should be re-used. If it is not that frequent, i.e., 0~50 percent hit ratio, then caching the output is not efficient or can even be useless.
- Acceptance level for data discrepancy: is the data inconsistent even though the two previous considerations are met?

These considerations are important to help decide if the cache command in the earlier PHP example should be used as-is or amended to be more relevant to the needs of the software and the speed of response required by users. When assessing the right caching strategy for your product, think about how the approaches in the next section can help guide your decision.

A top-down approach

The following diagram illustrates the components of a standard request. The upper layer forms the output of the lower layer. The further we travel down the layers, the less latency is required and the more frequently data is reused. Consider a caching strategy that:

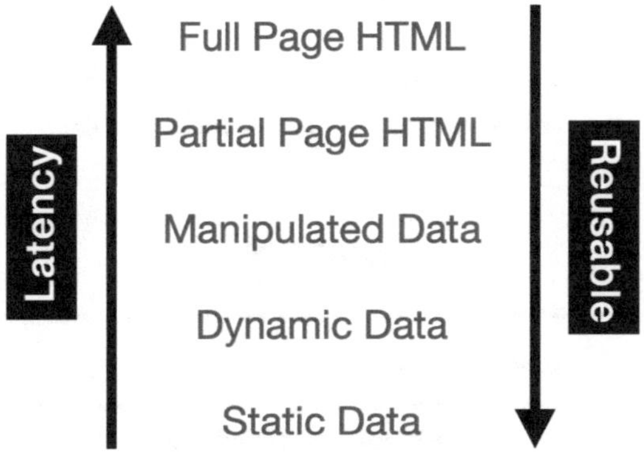

- Caches the top layers of the system to reduce the request's processing time.
- Duplicates properties of the lower layers in the upper ones, for example, bringing static data or shared data to the top.

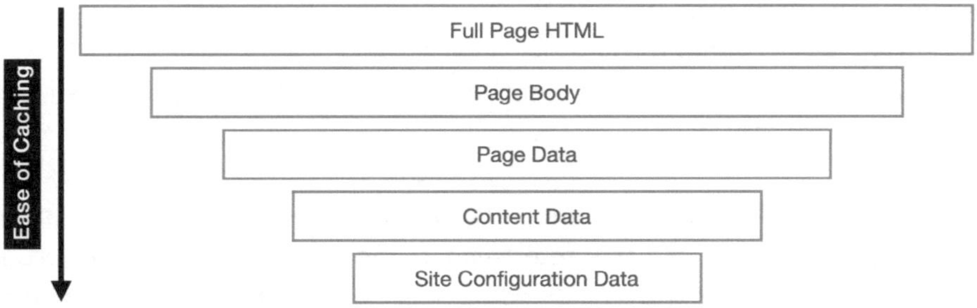

The further you travel down the layers in the diagram above, the simpler it is to cache data, resulting in faster request processing times. Going from the top down allows for the selection of specific layers to cache. Caching multiple layers could result in data duplication and is wasteful and energy-intensive as the data at the upper layers is already contained in the data at the lower layers. These factors must be considered to enhance the effectiveness and energy consumption of your caching strategy.

A left-to-right approach

Another method, shown in the diagram below, is to examine caching requests from a left-to-right perspective based on either generic or specific data characteristics:

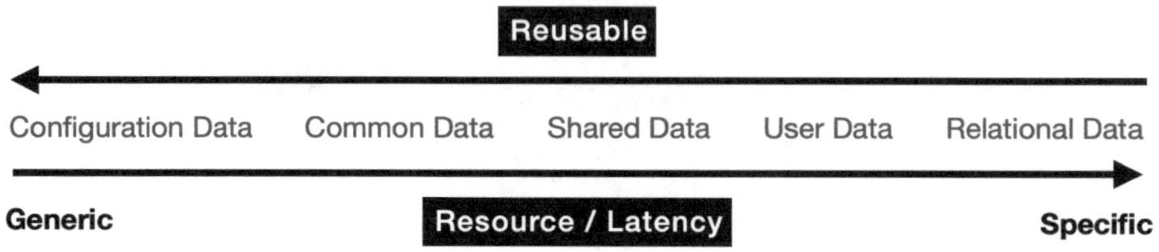

Here, the data to be cached is separated into different categories, for example:

- Header and footer (e.g., configuration data for an entire web page).
- Top-viewed articles (e.g., a sidebar)
- Main content (e.g., in the body)
- Current user information
- Information about users related to current articles (e.g., read, like, comment, etc.)

Typically, these data components are independent and can be cached separately based on their characteristics. Adopting a left-to-right approach enables you to determine which components require the most resources (i.e., those with the highest latency) and which components are most frequently used.

Also, consider the ease of caching. The diagram below helps to determine which components are most easily cached. It is useful to think about the following scenarios:

- If it takes 100 milliseconds to load the main body, which is reused in multiple places, then it is a candidate for caching.

- If it takes 120 milliseconds to load user information but it is not often required, e.g., most users are guests, then don't cache it.

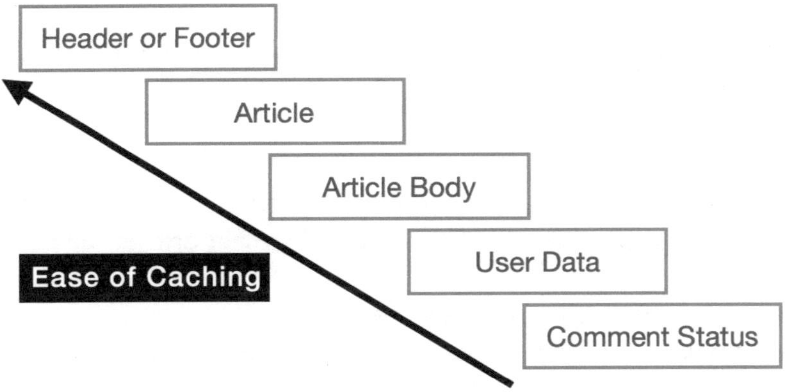

Data cache placement

For most back-end applications, there are two common places to cache data: (1) caching on a server with running code, and (2) caching remotely via a network. Let's consider the differences:

On server

On a server, there are two primary data caching mechanisms available: memory and file.

1. Memory (in-process): caching data as variables of the application process, e.g., array, hash map. Among its characteristics are:

- Saving as-is without serialisation
- Faster caching

- Suitable for applications written in NodeJS, Python, Java, .Net, with a main process and shared memory between requests
- Suitable for data with low latency, high access frequency, and small size (the latency of this type of data is often less than 1 millisecond)

2. Local file saving: data into a file on the server as a string, text, or binary. Characteristics include:

- Serialisation
- It depends on local disk speed, but usually fast
- For applications written in PHP, without the main process or non-shared memory
- Application running on VM, bare metal
- Early stage of the development (start caching) or later with a particular purpose
- Large data or native HTML, CSS, JS, images, and videos

Data caching on a server where code is deployed is typically efficient and fast. This is necessary for applications running in a few instances or as independent instances because a node's memory limits scalability and caching performance in terms of hit/miss ratio may degrade in distributed (multi-node) systems.

Remote caching

These are third-party solutions that cache data directly in memory. Here, common characteristics are:

- Available with predefined data types
- Network conditions have an impact on performance
- Suitable for a broad variety of applications
- Easy to scale without significantly affecting performance, independent of the number of code servers
- High hitting ratio as it shares cache with multiple code servers

- Suitable for small data items due to the RAM limit (and network bottlenecks for large data)
- Good for data with a low Time-To-Live (TTL) value

This technique is widely adopted as a cache solution because it has high performance and scalability. But be aware that even though it uses memory to store data, network latency can occur and affect performance.

Caching on remote disk storage is a technique that caches information either on a disk or in both memory and disk. Among its characteristics are:

- Generally available with more features and more supported data types
- Faster than a normal database, as optimised for SSD storage
- Easily scales in terms of resources or the number of nodes while maintaining performance at acceptable levels
- Good for large data or personalised data (e.g., recommendation, advertising, tracking, and analytics)
- Good for applications with high availability, resiliency, and auto-failover requirements
- Good for data with a high Time-To-Live (TTL) value

Selecting the right approach

The variables driving selection will be application-specific, but as a general rule, if your application is small and not very complex, selecting storage for caching may not be the right approach. Consider the following as part of your cache selection process:

- For interpreting languages with non-sharing memory like PHP, consider local file and network memory storage.
- For languages like NodeJS, Python, and Golang, consider memory and network memory storage.
- For small and common (highly requested) data, consider in-memory.

- For Java and .NET, consider in-memory to take advantage of the capability of handling complex data types as well as powerful server capabilities.

If your application needs to manage a high volume of traffic, you should exercise caution when deciding where to cache, as this will have energy and performance implications. Consider it from the perspective of data access: hot-cold aligns with local cache and even-distributed with remote cache:

- Hot means data that is accessed many times, and cold means the opposite, i.e., a node with hot data will be accessed more frequently and a node with cold data less so. So it may be better to do caching locally in memory and have it duplicated across processes.
- Even-distributed is used for data with a similar and consistently balanced access frequency. If cached at the local level, the hit/miss ratio will be reduced as data needs to be duplicated over servers, whereas a distributed remote cache will be more efficient due to its sharding and load balancing capabilities.

Cache duration

Assessing cache duration is important but can be a challenging aspect of decision-making because the duration of data caching has an immediate effect on discrepancies in data and performance. For example, optimising the cache Time to Live (TTL) value is more advantageous in some scenarios than deciding to cache in-memory.

To understand this better, consider the relationship between increasing cache TTL and cache performance in terms of hit ratio and storage cost, where:

- Increasing cache TTL will require more storage space.

- Increasing cache TTL will increase the hit ratio in a non-linear manner. It means that at some point, increasing cache TTL will no longer help to improve the cache hit ratio.

In practice, it is impossible to obtain an optimal cache TTL value without tuning the overall performance of the system. To achieve this, consider the following:

- Setting cache TTL to have a hit ratio as high as possible, i.e., at least 85%, and best if > 90%.
- Increasing cache TTL will impact storage resources; increasing the hit ratio may result in inefficiencies.
- Setting cache TTL with consideration for data characteristics, for example, data sensitivity and discrepancy.
- Using cache to throttle queries in the database, for example, 5–10 seconds of caching could be more than enough to avoid peak query loads in a database for hot data.

When considering cache duration, there is no one-size-fits-all solution, so carefully analyse the data characteristics and needs of your application.

Summary

The key to effective data caching is to carefully consider the performance and efficiency requirements of your application. Try to avoid cloning a standard caching strategy and, as a result, adopting an untuned level of performance and energy usage. You should actively monitor the cache hit/miss ratio as well as the storage size and use the results of this analysis to routinely optimise performance. By doing this, the energy performance efficiencies resulting from your actions can be significant.

3. Data exchanges

Unnecessary data exchanges that do not directly support a feature or function in the software can cause unnecessary network and storage load. It may also be necessary to translate the data that is being exchanged in order for the various components of a software system to function together. Such translations can be computationally expensive, for instance, when multiple translation steps take place instead of a single step from source to target. On top of this, the intermediate data formats can be excessively lengthy, resulting in the transmission of large files that consume even more energy in order to process them.

By simplifying formats, translation processes, and the number of data exchanges, it is possible to reduce energy consumption.

Data exchange selection considerations

Rarely are the reasons for selecting a data exchange method definitive, and it is often necessary to balance the advantages and disadvantages of a method against local and enterprise requirements. Again, there is no one-size-fits-all solution for data exchange. Choices need to be made, and the factors highlighted in this section should be evaluated as each will generate different levels of energy consumption and, consequently, carbon impact.

Data set characteristics

Data complexity

If the data entity to be transferred consists of multiple related elements or the specific components are unknown in advance, i.e., the required data elements vary in an ad hoc manner, then direct database access may be the most efficient option.

API-based data exchanges are now commonplace. Entity-based design is one of REST API's main design principles and brings the benefit of a predictable location for each entity (e.g., Plan 123 always lives at /plans/123), but it makes it more challenging to string together many related entities. Multiple API calls and coding may be required to reassemble the relationships between the various data elements. In certain instances, the use of an integration platform or enterprise service bus may help alleviate these issues.

Data formatting complexity is also an important factor to consider. For instance, flat files cannot readily represent hierarchical data because they are exactly that: 'flat'. JSON and XML are capable of representing more complex data models, but the REST architecture is designed to avoid complex query and result data, so a balance always needs to be achieved between architectural choices, programming language, and data complexity when selecting the optimal data exchange method.

Data set sizing

Transferring extremely large data sets frequently may require a file transfer or direct database connection in order to maintain performance. Although there are techniques to improve performance when transferring large data sets or high volumes of large messages via REST or an API, it is generally preferable to use methods that reduce the data set size in the first place.

Frequency of data updates

Processing and energy costs associated with replacing entire datasets via file transfer or direct database access can be substantial. This is only exacerbated if the data set is updated frequently. In order to avoid constant bulk resynchronisation, APIs and messaging system methods support transactional updates more efficiently and are therefore better options in these scenarios.

Data environment characteristics

Data flows

Always spend time to analyse the various planned and prospective data flows required by your application, as this will help in the selection of an optimal data exchange method. There are two common techniques that should be considered for both performance and energy efficiency:

- A Message Broker method employs middleware that mediates one-to-one, one-to-many, and many-to-one interactions. Based on the Message Broker pattern, extremely large and highly performant push-based fan-in/fan-out systems can be constructed. This is beneficial for asynchronous communication, unreliable networks, and applications involving large amounts of data.
- A Streaming Method should be considered when a large number of data sources continuously send data to a single receiving system, such as log and instrumentation data.

Frequency of data usage

If access to the most recent version of the data is required, a synchronous remote procedure call or API should be considered.

Data versioning

When a data provider has to provide various versions or schemas to support a range of consumer applications, an API may not be the optimal solution. An API is, ideally, a singular, consistent representation of a set of resources. Multiple schemas or versions within a single API are difficult to manage and frequently result in technical debt as well as creating processing load, which leads to higher energy consumption.

Data security

The tools and processes required to secure data at rest and during transmission are typically external to the chosen data exchange method. For instance, an API can be configured to require a key, whereas a direct database connection is constrained by database management system security controls, and web servers can be configured to protect data files. Although not directly related to the data exchange method, it is important to consider the necessary data security and controls when choosing between approaches.

Data transformation complexity

If the data exchange includes the need for significant processing and data transformation, particularly if the transformations are based on complex business rules, then a direct database connection using Extract-Transform-Load (ETL) tools is a preferred approach. This is especially important when the intention of the transfer is to move data from one resting location to another. Some API management platforms can also perform less extensive transformations, so the API method may be a practical choice depending on the nature of your application.

Connection persistence

Long-lived protocols are designed to maintain an open connection indefinitely, such as the Secure Shell Protocol (SSH). Persistent connections enable the streaming of data to end-user clients like web browsers and mobile devices. They are also valuable within the network when certain receivers, such as job processors, cannot be addressed. A good example is WebSockets, a protocol that can be used to facilitate bidirectional, persistent connections. Short-lived protocols are more transactional in nature, in which a specific action or succession of actions takes place before the connection is terminated. HTTP POST is an example of a short-lived protocol.

Consumer characteristics

Front-facing applications

End-user requirements are an important factor when determining the optimal data exchange protocol for your application. Consider the following:

- Text files, including the 'comma-delimited file' format, are readily readable by humans and can be used with common tools. If this type of direct usage is prevalent, then file transfers are often the best option.
- Developers and engineers typically use APIs to distribute text and hypermedia. If the receiving system is a web browser or equivalent front-facing agent, then a REST API is a good option.
- Systems exchanging private data and providing 'back-end' services are more likely to benefit from optimised Remote Procedure Call (RPC) methods than REST APIs.

Receiving system processes

Depending on the business processes supported by a receiving system, different types of data exchange methods will be preferable. For instance, the designers of a system geared towards batch processing of transactions may decide that all data transfers should be file-based, and while selecting an alternative data exchange method may be feasible, the balance between cost and benefit may not justify changing to a different method.

Usage by the receiving system

Where data is used to support a 'feature' and satisfies a specific requirement, such as a query to retrieve a set of attributes, an API will be the most suitable method. If, on the other hand, a large data set is being transferred and used as the basis for a 'platform' or reporting system, a file or database method will be more appropriate.

Summary

Data exchange method selection and use case suitability can result in efficiencies if done correctly, or conversely, result in significant processing, data transfer, and energy consumption costs if unsuitable approaches are chosen. The table below provides a summary of the use-case scenarios and their associated data exchange mechanisms. For your application, try to select the most energy-efficient option, as this will help reduce the amount of energy required for data exchanges.

Use Case Scenarios	Method Rank High | Medium | Low			
	Web Service	Messaging	Database	File
The data is required in multiple formats	H	M	M	L
The data is used in a client front-end	H	M	M	L
The data supports a feature	H	H	M	L
The data is frequently requested	H	H	M	L
The data changes frequently	H	H	M	L
The data is used in a back-end	L	H	H	M
The data has multiple flows and receivers	M	H	L	L
Assembling the data involves multiple entities and/or variable logic	L	L	H	L
The data is requested in multiple versions / schemas	L	L	H	L
The data set is very large	L	M	H	H
The data forms the basis of a larger platform	L	L	H	H
The data must be human readable	L	L	L	H

4. Data compression, aggregation & retention

Data that is not compressed or aggregated results in substantially higher levels of network and compute utilisation, each requiring greater energy and carbon costs. A business application must store transaction-based information, and after each business transaction has been completed, the majority of applications will retain transaction-related data for future reference. This information is frequently maintained 'alive' but unused. Year after year, it resides in memory or on a storage drive, consuming energy. By removing obsolete transactions from the application database, storage requirements can be kept from increasing, and energy can be conserved by compressing and aggregating only the data that is needed.

Data compression

The purpose of compression is to make the size of data smaller (whether a file, message, or any other format). Data compression can significantly reduce a file's storage space requirements. For instance, if we had a 20 MB file and were able to reduce it to 10 MB, we would have compressed it with a compression ratio of one to two, as the new file is half the size of the original file; or if we compressed the 100 MB file to 1 MB, it would have a compression ratio of one to one hundred, as the new file is one hundredth the size of the original file. The greater the compression ratio, the smaller the data size will become, and because the compressed data is smaller, it is possible to save money and time on storage and data transfers. Almost any form of file can be compressed, but it is important to follow best practices when selecting the right compression technique and method.

Data compression methods

There are two main types of data compression methods, commonly referred to as 'Lossless' and 'Lossy'. They are described as follows:

Lossless compression

This method allows a file to return to its original form when decompressed without losing data. Executables, as well as text and spreadsheet files, typically utilise lossless compression because the loss of words or numbers would detrimentally alter the information.

Lossy compression

Lossy compression loses data, while lossless compression keeps all data. With lossless compression, no data is lost, and techniques are based on encoding data in a more efficient manner. With lossy compression, the data is lost, so it is essential, when using this method, to differentiate between the data and the information that makes the data useful.

Data compression and energy consumption

Data compression is an effective method for saving energy because it reduces the quantity of data to be transmitted across a network. This is illustrated in the diagram below, which shows a data compression technique comprised of two algorithms, the first of which can take a given input (X) and generate an output (Xc) that can be represented in a smaller size (i.e., with fewer bits). The second algorithm operates on the compressed (Xc) data and reconstructs it when decompressed. Different data structures can be used to accomplish data compression; in this example, a 7-bit data stream is being reduced to a 5-bit data stream that can then be stored or transmitted over a network more efficiently.

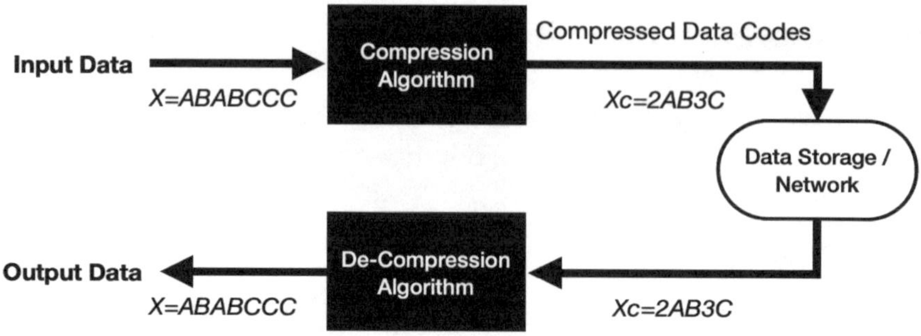

By eliminating redundant data, data compression reduces the amount of energy required to transmit compressed data. The diagram below illustrates the relationship between data aggregation, data compression, and routing algorithms.

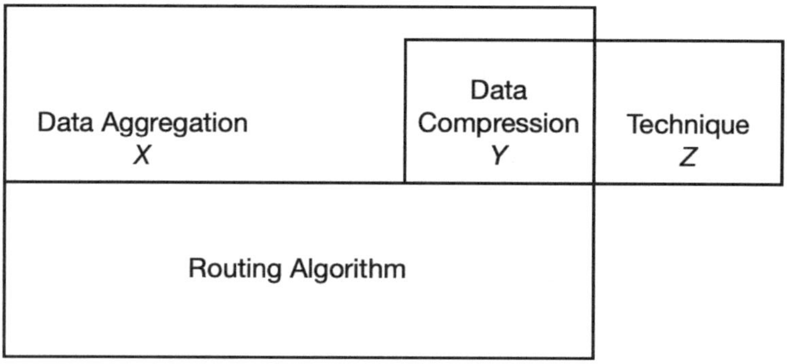

The process of data aggregation per X is positioned above the routing algorithm in the diagram above. The procedure involves extracting the minimum, maximum, and average values from aggregated data. This minimises data communication and is therefore more efficient. According to Y, the data aggregation technique could result in the loss of the original data's structure, which can be remedied by employing data compression techniques. As an extension of data aggregation, compression techniques distribute the compression algorithm across a network. The technique indicated by Z performs data compression independently at each local network node and does not require a routing protocol.

Data compression algorithms can therefore be categorised as local data compression algorithms, data aggregation algorithms, and distributed data compression algorithms. Using these methods of data compression, aggregation, and network-wide deployment can increase energy efficiency and reduce carbon emissions.

Data compression considerations

Modelling and coding are the two steps involved in the data compression process. Modelling is the initial stage in which data redundancy-related information is extracted and described as a model. A model is considered static if its parameters do not adapt to shifting data characteristics. In contrast, it is adaptive if parameter adjustments are made based on the changing characteristics of the data.

In the coding phase, the model is compared to the data, and differences are encoded in binary. This difference is referred to as the 'residual' and can be encoded in a sequence using fewer bits than the original data.

Some applications cannot tolerate a certain amount of latency or data loss, so consider the following data compression factors to help determine which are most important for your application:

Complexity in processing and memory capacity

Limitations resulting from network processing speeds require the development of less complex compression algorithms with smaller code sizes. It is always preferable to design algorithms that perform the majority of data processing at the sink as opposed to compressing data at individual network nodes. This has the dual benefit of requiring less complex processing algorithms that also reduce both processing costs and power consumption.

Data exchange

During data transmission and reception, a large amount of energy is consumed. Some compression algorithms can remove redundant network-communicated data to improve network efficiency; however, these algorithms need to perform additional processing to reduce data transmission, and as a consequence, both the transmitter and receiver are required to perform increased data processing, resulting in additional energy use. Different variables like distance, frequency, and data volumes need to be balanced when making choices.

Redundant data acquisition

It is possible for duplicate or redundant data to be produced during data collection, transmission, and storage. More sophisticated compression algorithms can identify and use redundancy techniques to improve efficiency.

Reliability and scalability

The measurement and spatial redundancy of data across a network can be used to enhance data reliability during data communication. Data compression algorithms that can scale and grow in proportion to the network's expansion should be used.

Compression routes

Under normal circumstances, data is compressed at the source and decompressed at the destination. Data is also made accessible and forwarded across network nodes, and compression can occur on and along network routes to support data processing. The number of on-route compression and decompression operations will directly influence the overall efficiency and energy utilisation of the data transfer.

Robustness

Compression techniques should be designed to function properly even in the event of a network failure. This requires a compromise to be made between energy efficiency and robustness, as additional deployments may be needed to increase resilience and accommodate potential failures.

Security

Some applications require a specific level of security, which may conflict with data compression. Some security protocols may necessitate data encryption prior to data transmission, with data decryption occurring only at the sink or destination. Algorithms for data compression and security protocols should be developed with one another in mind in order to maximise efficiency.

Real-Time

Some time-sensitive applications require that compression be performed on a sample-by-sample basis, thereby limiting the compression ratio. When implementing an algorithm for applications that necessitate real-time operations, the algorithm's performance needs to be considered in order to maximise energy efficiency.

Quality of information and service (QoI or QoS)

Quality of Information (QoI) or Quality of Service (QoS) refers to the quantity of the data at the destination and its level of resemblance to the original data. This requirement needs to be evaluated in the context of the 'Lossy' and 'Lossless' compression methods described previously to determine the level of data integrity needed by the software and/or end user. All of these considerations contribute to the overall energy performance of the digital product or service.

Data aggregation

Data aggregation is the process of collecting data and summarising it for presentation. The accuracy of data analysis and the delivery of results are dependent on the quantity and quality of the data collected. Data aggregation is frequently used to provide statistical analysis for compiled research data and to produce summarised data. The collection, curation, and presentation of data in a summary format is commonplace, so this technique is widely adopted.

During data aggregation across a network, data packets are combined using various routing strategies. This is accomplished by manipulating the derived characteristics and statistics of data sets and then sending them to the sink. Look again at the diagram in the earlier data compression section (p.124), showing how routing algorithms and data compression techniques are needed for efficient data aggregation. An efficient data compression algorithm can reduce the data's size and use fewer resources to compress the data. But note that user characteristics and needs should be taken into account when making compression decisions, as irrelevant data can be eliminated and improve processing efficiency.

Data aggregation techniques and types

Data aggregation can be achieved through a variety of methods. Despite it being typically a straightforward process of combining data values across large data sets, there are different factors that need to be considered based on the nature of your application:

In-network aggregation

This is a general process of gathering and routing information through a multi-hop network (i.e., where a network's overall coverage is greater than the range of a single node).

Tree-based approach

The tree-based approach is based on the construction of an aggregation tree. Visualise a tree structure in which the sink node is the root and the source nodes are the leaves, where data-derived information flows from the leaves (source) to the root (sink).

Cluster-based approach

A cluster-based method is used to aggregate larger data sets across the entire network. It is subdivided into clusters, each of which has a cluster head chosen from cluster members.

Multi-path approach

In a multi-path approach, partially aggregated data is sent to a single parent node that serves as the 'root node' of the aggregation tree. A node can send data along multiple paths, and each node can transmit data packets to a range of inputs.

Time vs. spatial data aggregation

Spatial aggregation and time aggregation are the two most commonly used types of time-based aggregation. Spatial aggregation takes data points from different sources within a given time frame. An example of spatial aggregation might be the calculation of the total number of prospects across all marketing channels. Whereas, time aggregation involves collecting data from a single source over an extended period of time, such as calculating the number of leads from a single marketing channel per month or per year.

Manual vs. automated data aggregation

Data aggregation can be performed manually when there are a limited number of data sources or a small volume of data, but it is preferable to automate whenever possible. For instance, downloading CSV-formatted reports and uploading them to a dashboard takes time. Automation has the dual benefits of reducing the likelihood of human error and saving time.

Real-time data aggregation

Real-time data aggregation provides immediate updates to dashboards or live reporting services. However, using real-time data aggregation for all analytics is not always feasible or efficient. In most cases, platforms offering real-time data aggregation can be set to provide hourly updates, as this saves on processing and network traffic and reduces energy usage.

Data aggregation vs. data mining

Data mining is a considerably more complex and technically involved process than data aggregation. Typically, data mining is used to discover trends in large data sets and sometimes requires machine learning to cater to more advanced use cases. Data mining is also used to make predictions or educated estimates about user or system behaviour based on data. Data aggregation is a much simpler method of data analysis and is most effective when applied to simpler data sets from multiple sources requiring categorisation.

Data retention

Data policy governs data retention more than technical or code requirements. Nonetheless, it is important to recognise that data retention approaches can be complex and challenging, often necessitating a granular, planned, and flexible approach. Some data may only be required for a brief time, such as cookie session data, which should be deleted promptly upon the expiration of the browsing session. However, other information, such as employment records, may be required for a considerably longer period of time. The policy and duration of data retention will have an impact on data storage requirements, retrieval and transmission across networks, and the need for data compression solutions, all of which have a significant impact on energy consumption and, consequently, carbon costs.

You should consider the following factors when making your data retention choices:

Understand the risks

Keeping personal data for too long or not long enough could pose legal, security, business, and reputational risks.

Data purpose

Understanding the personal data being processed and its range of uses is important. Some companies refer to these as Records of Processing Activities (RoPA). Consider creating a central record of data processing activities by mapping data and their associated flows.

Retention periods

Certain situations have legal retention requirements, such as the retention of employment data. Where laws do not specify how long data should be retained, it is wise to investigate further and make informed decisions regarding how long the data may be required and, consequently, what the retention period should be. This will impact the amount of data storage required by your application, with resultant costs, both financial and carbon-related.

Controllers, processors, and sub-processors

The majority of businesses outsource a portion of their data processing to third-party providers (also referred to as 'processors'). As a controller, you must inform processors of the length of time they must retain personal data and ensure that this is covered in contracts with your vendors.

Data retention policy and schedule

A data retention policy should outline the compliance requirements, and the schedule should confirm the specific retention periods required for the data.

At the end of the retention period, put a step in to check whether there is a need to retain the data for longer before deleting it.

Summary

In a digital world where data plays such a crucial role in the successful operation of products and services, it is critical to employ appropriate data management techniques. This section has shown that data compression, aggregation, and retention approaches and techniques are available to effectively manage these situations, and if appropriate methods are chosen, they can help to improve the performance and efficiency of applications. In turn, this can increase energy efficiency and decrease the carbon footprint of data processing, transmission, and storage.

5. Image sizing

As images and video become higher-definition, they are increasing in size, resulting in increased network load and energy demands. Reducing image resolution as well as the total number of images can have a substantial effect on computational processing and network utilisation. Image-intensive applications that operate in a web or mobile browser can benefit from adopting the best practices highlighted in this section.

Low-carbon images

According to the HTTP Archive[82], images are the most widely used resource type on the internet. As a direct result of the growing usage of images on websites, web pages are becoming increasingly data-intensive.

---

82 Source: httparchive.org 'Report: State of Images'

As we discover more about the web's environmental impact, we have a greater responsibility to use digital resources carefully. Fortunately, there are techniques for reducing the weight of images, ranging from basic actions such as 'image optimisation' to more sophisticated enhancements such as 'lazy loading'.

Key techniques

Sustainable web design, like user experience (UX) and service design, should always begin with a thorough understanding of the user's requirements. This information is core to understanding whether an image or video is useful and its importance to the user journey being undertaken.

From the perspective of a sustainable engineering purist, the ideal low-carbon image is one that does not exist, as it would have zero carbon cost. Rather than considering sophisticated image compression or data optimisation processes, it is preferable to avoid adding functionally unnecessary components, and images and video should be considered in the same way. Only using them when they significantly improve the user experience would be an easy way to start improving the energy efficiency of an application.

When an image is really required, there are two straightforward ways to reduce its impact in terms of energy consumption and carbon emissions:

Limit image quantity

Using a single, high-quality image as opposed to multiples is one approach. When creating a website page that introduces a company's team members, for instance, displaying everyone in one photo might also be viewed as a more sociable alternative than presenting each face separately. Another option is to display no images by default and request that the user 'enable' image display. The example below demonstrates how a website might implement this technique to reduce its carbon footprint each time the page is loaded and displayed. You can immediately see how simple and lightweight the page becomes as a result.

Reduce image size

A good web design allows for the display of images without the need for large file sizes. Images with a full width also influence the page's weight. Try to resize images outside the browser and avoid uploading large images to display small ones. Increasingly, consumers and end users are recognising the energy-related cost of highly graphic and visually rich websites and beginning to make conscious purchasing decisions and brand associations based on digital sustainability.

Alternative Approaches

Sometimes it is necessary to have a visual representation as part of the user experience or customer journey. Consider the following techniques (that will be explained further under 'Green Design' in Part Three) that can help reduce energy consumption and carbon emissions in these situations:

Use SVG format

When an image consists solely of vector graphics (lines, points, and polygons), the SVG format (Scalable Vector Graphics) is a good choice. SVGs are typically smaller in size and can be embedded in code to eliminate the need for a server request. They can be interacted with through CSS (Cascading Style Sheets) and JS (JavaScript) and are therefore readily accessible and straightforward to integrate into most new applications.

ASCII, Emoji, CSS Pattern, Web Font

If the design is basic, CSS patterns, emojis, or even ASCII (American Standard Code for Information Interchange) art can be substituted for an image. They can also be multiplied to create patterns with the illusion of greater complexity. When an image contains only text, a standard web font should be considered.

Image optimisations and formats

When an image is required, select a format that is as energy-efficient as possible. The following table provides a summary of the optimal format for the main image types.

The choice of which format to use depends on the image type needed for each use case. In summary:

Image Type	JPEG	PNG	SVG	WebP / AVIF
Photographic Image	X			X
Simple Illustration or Logo			X	
Complex Illustration or Logo, Picture w/ Text or Lines & Text		X		X
Support for Old Browsers	X	X	X	

- JPEGs are universally recognised, have high levels of compatibility, and are more efficient for displaying photographic content.
- PNGs have transparent backgrounds and are better suited when detail is required (brilliant colours, lines, and text); for example, they are the preferred format for illustrations, typography, and logos. PNGs offer lossless compression, so they maintain image quality better than JPEGs.
- SVGs are sometimes preferable to PNGs, particularly for icons and images with limited colours, but sometimes compression tools can obtain the best results.
- The WebP format has the benefit of being compatible with many browsers and is considered to have high compression efficiency compared to JPEG and PNG.
- The AV1 Image File Format (AVIF) is a specification for an image file format used to store images or image sequences encoded with AV1 in the HEIF (High Efficiency Image Format) format. It competes with the HEIC (High Efficiency Image Container) format, which employs the same ISOBMFF-based (ISO base media file format) container format but with HEVC (High Efficiency Video Coding) compression.

It is important to note that these new formats are compatible with modern browsers and are designed for users with higher-powered devices, which may be a limitation that should be taken into account when making a selection. It is essential to include a fallback solution to support older devices with these approaches.

Compression process

Consider employing manual compression on a website with a small number of images, as an API may be too complex and expensive. If there are numerous images, it may be preferable to compress them in bulk. A library such as ImageMagick is a tool for batch-resizing images. Nowadays, many Content Management Systems (CMS) offer built-in compression features or image compression plug-ins.

Image effect

Consider using image effects that reduce file size, such as dithering images to display fewer colours. Keep in mind that not all of the available effects have a positive impact on image dimensions. Try to make choices based on your targeted page weight, user requirements, and overall web design.

Remove metadata

Removing an image's metadata to reduce its file size is an effective method for improving performance and reducing energy use. Metadata can be effectively removed using readily available tools (e.g., ExifTool, ExifCleaner, Metadata Cleaner, etc.).

Image integration and responsive design

Use a modern approach to image integration that enables you to adjust the size and weight of the images based on the device type of the target user. When doing so, consider the following:

- Use the.img tag with srcset to specify multiple image sizes.
- Use the picture tag in conjunction with srcset if you require multiple image formats.

- Use resources such as MDN (Mozilla Developer Network) Web Docs, where you can find comprehensive guides for integrating responsive images. However, always offer a fallback for browsers that do not support image, srcset, or the alt tag.

Lazy load

Lazy loading is a technique that waits to load certain parts of a web page, especially images, only when they are needed. Instead of loading everything all at once, lazy load will load images below the viewport of a device. On a modern browser, it is possible to use the native lazy loading feature, and using a lazy loading library can enhance the loading time for all users on pages with many images. However, if you use a JavaScript-based library, you should provide a non-script alternative for users who disable JavaScript; otherwise, things may not load at all.

Base64

Instead of linking to an external file, it is possible to embed base64-encoded images within a document. You can simply convert the images with an online tool and integrate them into HTML or CSS.

Use a content delivery network (CDN)

In the case of complex website architecture or for a large or global user base, a Content Delivery Network (CDN) can be an effective means of reducing response times and latency. A CDN can handle image compression and optimisation, but like all third-party services, consider the dependencies and privacy risks that might be created.

Use Cache

A well-configured cache is essential for reducing page weights and image sizes when a user navigates between pages.

Summary

Creating and utilising low-carbon images and the various techniques to minimise impact through compression or choices to reduce the number of images can play a crucial role in enhancing the energy efficiency of an application, especially websites. This is a popular topic, and I've only covered an introduction to the main techniques here. There is a lot more literature, guidance, and support available on the internet if you want to explore it further.

6. Unused features

Unused software features need to be removed in order to eliminate their energy consumption. Adopting the principle of always removing anything that is unnecessary for the end user or needed to secure a product's licence to operate is a highly effective decarbonisation strategy.

For every feature that needs to be developed, it must be demonstrated that it is genuinely useful, either by enhancing the user experience or by making the user more productive in their tasks.

A product perspective

Product managers frequently prioritise and focus on developing the next new feature or piece of product functionality in order to provide end users with the best possible experience. Product managers should also retire features that are no longer being used or contributing value to the product. Software engineers can support product managers in identifying which features are not being called or used by viewing usage data at the code, component, and function level.

For the product manager and software engineer to make effective decisions, they must use data to gain a deeper understanding of user behaviour and sentiment. When usage is extremely low or an outdated feature receives repetitive negative feedback, product managers can use this information to make the case for the removal of the feature.

To build a data-driven approach to removing product features, the following should be considered:

Product usage

The first step is to conduct a product analysis. Gathering information that helps identify a feature that is underutilised qualifies it for removal.

Building a trend analysis that looks at in-depth feature-level interaction over a period of time (30–90 days is a good proxy) and identifying whether usage is intermittent or infrequent helps to create a shortlist for continued review. Not every feature (or product) is intended to be used daily, weekly, or even monthly, so it is important to consider expected usage patterns and trend data to determine a product's true usage.

Engagement

The greater the amount of data available to support the removal of a feature, the greater the likelihood that it will be accepted by all stakeholders. The following considerations are important:

- Examine the time customers and end-users spend on specific application pages or the frequency with which specific actions are completed.
- Examine the routes users are taking to access the feature and identify any barriers.
- Identify additional features or even workflows in the application that can more efficiently address a user's needs.
- Examine the data over time to determine if there are patterns of declining end-user engagement, and then determine the cause.

User segmentation

User segmentation is one of the most effective methods for understanding how end consumers interact with various product features. When examining which user segments access specific features, there are several points that should be considered:

- Company size
- Persona
- Net Promoter Score (NPS) response
- Role

Qualitative data

Equally important to the quantitative data captured during product analysis is the assessment of qualitative data, particularly customer feedback. Trends indicating a decline in feature usage require an understanding of user engagement and behaviour. Customers can provide some of the most valuable insights available since they are the ones using the product and its features.

Impact monitoring

Once a decision has been made to retire a feature, it is important to monitor the repercussions of removing it, particularly for any users of the feature that was removed. It is good practice to continue monitoring usage, sentiment, and feedback after a product feature has been removed.

After collecting the data, consider the following:

- Identify alternate pathways users are taking to accomplish the same task or workflow as the retired feature
- Whether alternate features are seeing increased usage
- Customer feedback concerning the retired feature
- NPS movements post-feature retirement

A technical perspective

As software applications expand, they develop underused areas of code; identifying and removing this 'redundant' code allows for a more streamlined customer experience and can increase future engineering productivity. The following areas are important to consider:

Unused code

Understanding the potential causes of unused code in a software application can reveal important data points. For example, look for:

- Feature flags that didn't go live
- Code that customers didn't discover
- Code that customers didn't find useful
- Low-quality features
- Features that didn't get traction with customers

- 'One-shot' features that were used at a point in time but don't hold future relevance

Taking action on unused code

After locating unused code, there are two options available: either enhance it and ensure that it is suitable for its intended purpose, or remove it.

Improve it

Review the original feature specification and conduct a gap analysis to identify the missing or deficient areas. Once the feature has been understood and its end-user requirements have been validated, move the code changes necessary to enhance it to the backlog.

Remove it

Removing superfluous code reduces a software system's size and complexity. This can result in a faster build and deployment cycle, a reduced time to checkout new code, and an increase in the application's overall energy efficiency.

Removing 'Deep Code'

It is relatively simple to remove controller endpoints and UI elements, but when code endpoints invoke other active features, the task becomes more complicated. Several proactive and reactive techniques can be used to reduce the risks associated with removing 'deep code':

Proactive

- Once a controller endpoint is removed, tooling is available to support many languages and IDEs (Integrated Development Environments), which can help identify unused code.

- An extensive automation test suite can be used to identify any code interdependencies.
- Instrument the code more deeply to determine if it is really unused. Telemetry at the code block level can be used to record its usage, take measurements, and support more informed decisions.

Reactive

- Slowly and in a structured way, remove unused code, increasing the opportunity to catch any issues that may arise from its removal.
- Include the removal of features in release notes to allow customers and end users an opportunity to provide feedback.

Summary

Over time, features and elements of code will inevitably become obsolete. Unused code slows down software development. If this code is part of the user interface, it can also increase the cognitive burden and degrade the user experience. Through instrumentation and analysis, it is possible to identify unused code and make an informed decision on whether to enhance it or remove it. Where removal is possible, complete the task as quickly as possible to maximise energy efficiency gains and carbon reduction.

7. Logging, looping and polling

During software development, logging the application's processing and test results is needed for identifying bugs and addressing performance issues. However, writing this data to disk and storing sizeable log files consumes energy. When the final application is deployed in a production environment, a significant portion of this logged data is not used. By following a framework that reduces the level of logging when no logs are needed, excessive energy consumption can be avoided. You should try to detect and eliminate loops that can never accomplish their intended purpose, such as polling an unreachable server. On a regular basis, identifying and eliminating unnecessary logging, looping, and polling can contribute significantly to the performance and energy efficiency of software.

Logging

Logging is an essential component of software architecture, allowing for the identification, debugging, and management of unusual events in production environments. It is highly likely that hundreds of gigabytes of log data are being generated and stored in any production operation of significant size. Some of these records are monitored and queried as part of incident investigations, but the vast majority of log data is never used, which not only increases costs but also consumes energy and has a carbon footprint.

The remainder of this section covers techniques and practices that can help improve your logging efficiency.

Optimising and enhancing log data

Adopting best practices for optimising and enhancing log data, such as data compression and log extraction, are effective strategies to consider:

Data compression

In line with Technique 4: Data Compression, Aggregation, and Retention, employing data compression methods and algorithms to optimise and enhance log data can reduce the overall storage and transfer costs associated with network bandwidth consumption.

Data parsing

Combine parsing and indexing techniques to manage large log volumes. Parsing is an effective technique for understanding data sources, processing events, and storage methods. Logs are transformed into data fields that are simpler to index, query, and store after parsing. Numerous monitoring solutions include tools for parsing logs and collecting key-value pairs based on standard delimiters like colons or equal characters.

When thinking about parsing, there are a few important factors that need to be considered to successfully manage multi-line log statements:

- Data replacement
- Setting rules to replace content at index time when certain conditions are met
- Discarding unnecessary data
- Setting rules to discard logs when log messages become redundant

Metadata parsing

Parsing can be used to derive useful information from log metadata. For instance, cloud infrastructure can extract the metatags of default cloud providers in order to enrich the data with additional context.

Log transfer reliability

Server connectivity issues can result in missing logs, which can compromise log integrity. To safeguard the reliability and integrity of logs, log monitoring solutions should preserve logs. For example, some Application Performance Management (APM) solutions use the advanced log transfer protocol (ALTP) to maintain data persistence.

Deduplication

Duplicate logs are one of the most challenging aspects of log management. Log deduplication can be prevented by setting parsing rules, helping to enhance log optimisation and quality.

Logging frameworks

Logging frameworks provide an abstraction layer so that developers can use an API more easily. As an illustration, the following logging statements associated with shopping cart management are frequently used in the Java SLF4J framework:

```
LOGGER.debug("Customer put item {} in the cart", item.getId());
LOGGER.debug("Item {} price is {}", item.getId(), item.getPrice());
LOGGER.debug("Cart price is now {}", cart.getTotalPrice());
```

Most frameworks operate in a similar way, with the following principles:

- Each logging statement makes use of a logging level
- The configuration file states a logging level
- If the criticality of the statement's logging level equals or is above the configured criticality, the log is written; otherwise, it is not

Log statements

Continuing with the example above, even if a log is not written because the configured criticality is greater than the statement's criticality, the call will still be executed. 'cart.getTotalPrice()' is executed in every case in the following example statement:

```
LOGGER.debug("Cart price is now {}", cart.getTotalPrice());
```

Consider a scenario in which the calculation of the cart's total price necessitates more complex logic, such as calling an external service to apply a discount. Therefore, regardless of whether logging occurs due to the 'debug' level, time will be wasted calculating the price of the cart.

There are two ways to solve this:

- Guard the statement with an 'if' statement depending on the log level, for example:

```
if (LOGGER.isDebugEnabled()) {
  LOGGER.debug("Cart price is now {}", cart.getTotalPrice());
}
```

- Or use a lambda expression in the statement. Lambdas are 'lazily' evaluated, which means the cart price will only be computed if the logging level is 'debug', for example:

```
LOGGER.debug(formatter -> formatter.format("Cart price is now {}",
cart.getTotalPrice()));
```

Adding a guard condition can increase complexity, so it is essential to ensure that the logging level of the guard matches that of the statement. In all cases, the abstraction layer has resulted in additional processing, and you must be aware of this in order to prevent logging statements from negatively impacting the performance of your application and thereby creating energy inefficiencies.

Blocking logging

Another performance-related factor is that by default, logging is 'blocking'; that is, the runtime executes the log statement and writes the log to disk. The rate of disk writing is contingent on the performance of the hardware. If the process of writing is slow and prevents the application from proceeding because it is blocking, then logging has effectively become the application's bottleneck.

In logging frameworks, appenders are components that are entrusted with writing. Appenders hide behind the API and are configured independently, allowing them to be modified without requiring a code update. Some logging implementations, such as Log4J or Logback, offer asynchronous appends to prevent blocking, resulting in improved performance.

Async appenders can be implemented in a variety of ways, such as Logback, which buffers log messages in a queue. A specialised thread retrieves these messages from the queue and writes the associated message. The following configuration options should be considered:

- Queue size
- Logging the level of messages and identifying those to discard when the queue starts filling up
- Setting a queue capacity threshold to start the discard process

Performance of log metadata

In addition to the log message, log files typically include metadata to enhance their usefulness. Occasionally, the metadata may be larger than the log message. For instance, consider the following log message and metadata from a Spring Boot application:

```
2018-12-17 13:56:54.906  INFO 1 --- [  restartedMain]
c.e.configmgmt.demo.DemoApplication      : Started DemoApplication in
3.833 seconds (JVM running for 4.303)
2018-12-17 13:57:05.595  INFO 1 --- [nio-8080-exec-1] o.a.c.c.C.[Tomcat].
[localhost].[/]       : Initializing Spring DispatcherServlet 'dispatcherServlet'
2018-12-17 13:57:05.595  INFO 1 --- [nio-8080-exec-1]
o.s.web.servlet.DispatcherServlet       : Initializing Servlet 'dispatcherServlet'
2018-12-17 13:57:05.603  INFO 1 --- [nio-8080-exec-1]
o.s.web.servlet.DispatcherServlet       : Completed initialization in 7 ms
2018-12-17 14:40:57.731  INFO 1 --- [      Thread-9]
o.s.s.concurrent.ThreadPoolTaskExecutor  : Shutting down ExecutorService
'applicationTaskExecutor'
```

The log message is to the far right, while the metadata is to the left. Here, the metadata contains:

- The timestamp, down to the millisecond

- The log level
- The thread name
- The fully-qualified Java class (abridged)

Other possible examples of useful metadata that can be written in logs are:

- Any state-related metadata
- Source IP/hostname of the coming HTTP request
- The method the log message originated from
- The source file the log message originated from

Some of these options will have an impact on performance because the logging framework must compute them, so you should only use them if they are actually required. If the metadata does need to be included in the final log message, the trade-off between use and performance needs to be carefully evaluated beforehand.

Log aggregation

Individual server logs are typically not very useful for production purposes. To investigate production issues, it is better to have a comprehensive view across multiple servers. Log aggregation is essential in these circumstances and requires the use of a centralised logging service to aggregate logs from all relevant sources. Common solutions for log aggregation include Elastic Stack, Splunk, and Graylog.

Scraping log files is a common method for moving records into an aggregation service. In the Elastic Stack architecture, for instance, the Filebeat component 'watches' for log files and transmits scraped content to ElasticSearch for use.

Scraped metadata

Scraping can also include metadata. Consider an application cluster deployed on multiple nodes, each of which writes to its own log file. When log files are aggregated in the centralised service, there must be a way to identify the host that generated the logs. There are two methods to accomplish this: (1) the logging framework accesses the hostname and writes it into the log file, or (2) the scraping agent adds the hostname metadata to each log message. The second approach is generally more efficient.

When the scraping agent adds the metadata, a number of additional benefits can be derived:

- Ability to distinguish between different log files on the same host
- Ability to set the source environment, e.g., production, staging, and development
- Ability to add any kind of additional metadata that is useful when analysing production issues, such as the cloud zone

Log read-write optimisation

There are two main ways to implement search on log data: (1) storing the data in a structured way so that queries can run faster, though the storing itself can be slower, and (2) storing the data as it is, in order to make storing faster but querying slower as a trade-off. These methods are known as 'schema-on-write' versus 'schema-on-read'.

For quicker querying, logs must be stored in a structured manner. Since a log message is essentially a string, the log aggregation service must parse, structure, and digest it. Typically, a tool is used to parse log messages, which accepts data from multiple sources, transforms the data, and then transmits it to the aggregation service.

There are a variety of techniques, tools, and strategies for managing the logging overhead of an application. The range of factors and parameters that influence decisions for any particular application makes this a relatively complex challenge for the software engineer. However, the time invested is worthwhile and can enhance the overall effectiveness and efficiency of the software product being designed and developed. And with these efficiencies come opportunities for energy reduction and, consequently, lower carbon impact.

Looping

Identify and removing loops

Before a loop can be eliminated, it must be identified. A typical technique for detecting a loop in a linked list (a sequence of data structures connected by links) is to:

- Traverse the list individually and put the node addresses in a Hash table
- At any point, if NULL is reached, then return FALSE
- If the next of the current nodes points to any of the previously stored nodes in Hash, then return TRUE

To remove the loop, get a pointer to its final node. For instance, a node with the value 5 in the diagram below. Once a pointer is added to the last node, it is possible to make the next node NULL, and the loop is removed.

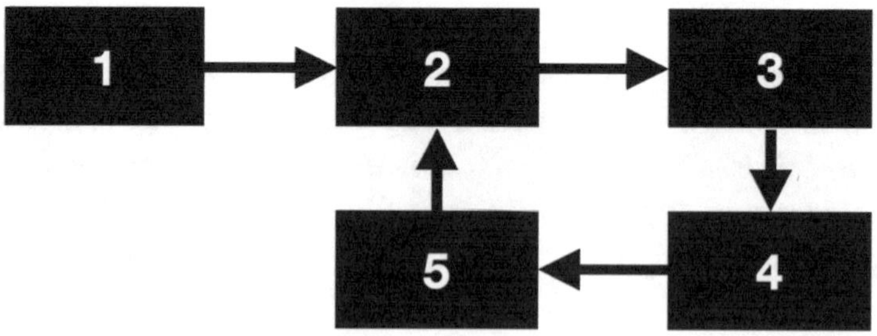

Hashing or Visited node techniques can be used to get the pointer to the final node, which is the very first node whose next node has already been visited (or hashed). The Floyd Cycle Detection Algorithm[83] can be used to detect and eliminate the loop. Slow and rapid pointers converge at the loop node in Floyd's Algorithm. This node can be used to remove the loop.

When Floyd's Algorithm is used for loop detection, there are a number of methods for removing the loop:

Method 1

Floyd's Algorithm for detecting cycles terminates when the fast and slow pointers intersect. This shared point is one of the loop nodes (2, 3, 4, or 5 in the previous diagram). Consider carrying out the following activities:

- Store the address of this in a pointer variable.
- After that start from the head of the Linked List and check for nodes one by one if they are reachable from the variable.
- Whenever a node is reachable, the node is then known to be the starting node of the loop in the Linked List, and a pointer to the previous node can occur.

Method 2

This second method is a variation, where:

- Start by detecting the loop using Floyd's Cycle Detection Algorithm and setting the pointer to a loop node.
- Count the number of nodes in the loop.
- Fix one pointer to the head and another to a selected node from the head.
- Move both pointers at the same pace; they will meet at the loop starting node.
- Get a pointer to the last node of the loop and make the next = NULL.

---

83 Find out more on Floyd's Cycle Detection Algorithm at geeksforgeeks.org

Method 3

This approach is based on optimising Method 2. Here, it is not necessary to count the number of loop nodes. Start the slow pointer from the head and move both the slow and fast pointers at the same speed until the slow and fast pointers do not meet; they will only meet at the beginning of the cycle.

Method 4

A method built on hashing. Hash the addresses of the linked list nodes in an unordered map and verify if the element already exists in the map. If it exists, a node that already exists in the cycle has been reached, so the next pointer of the last node should be set to NULL.

At a software engineering level, looping can cause serious performance issues and burn energy continuously if not addressed, so it is important that you spend time considering which method might be best suited to improve the efficiency of your software product.

Polling

Avoid polling

Polling is a technique used to improve responsiveness where a CPU enters a short loop, checking a condition repeatedly until it becomes true. Polling is almost always inefficient and increases power consumption as it forces the CPU to be constantly active.

As polling consumes CPU time and can steal it from other applications in a multitasking environment, it should be discouraged. This is particularly true on mobile platforms, where AJAX-style applications routinely poll a web server for new events. Instead, consider utilising the Extensible Messaging and Presence Protocol (XMPP), a WebSocket, or a custom protocol based on the application's specific requirements when replacing polling on mobile devices over a network with push notifications.

Polling should be avoided whenever possible, as it causes unnecessary CPU cycles. Not only is wasting CPU cycles inefficient and degrades performance, but it also increases energy consumption and can have a substantial carbon impact.

Interrupt

A special instruction or an exceptional processor condition can result in a software interrupt. A software interrupt is one of the ways to communicate with the kernel (the main interface between a computer's hardware and software), particularly when handling errors.

When deciding whether to poll or interrupt, you must consider the nature of the event you're attempting to track and the most appropriate response to it. When nothing is happening, interruptions require no processing, but when something is occurring, they demand full attention. These situations are not optimal if the interrupting event is external and has 'noisy edges' or 'rapid pulses' (i.e., is inconsistent in nature).

When configuring your system, consider the following:

(1) Use a handler to manage an interrupt when the event of interest is:

- Asynchronous
- Urgent
- Infrequent

(2) Polling might be a better fit if the event of interest is:

- Synchronous
- Not Urgent (i.e., a slow polling interval has no detrimental effects)
- Frequent (i.e., the majority of polling cycles create a 'hit')

When setting up interrupts in either the background routine or for a higher priority than the code runtime, care must be taken. If an interrupt flag is already set prior to it being enabled, the interrupt routine may be called incorrectly as soon as it is enabled. The forcing of the interrupt makes the system work in the same state, for example:

```
Set the edge to detect the next interrupt event
Clear the interrupt flag
if (the event has already occurred)
{
  Set the interrupt flag to force the interrupt
}
Enable the interrupt
```

Also, be mindful of the following factors when choosing between interrupts and polling:

- If the time of the response to an event has to be consistent or if the time of the response has to be within a certain timeframe, then selecting an interrupt may be a good choice.
- Consider threading, i.e., where two pieces of code access the same data at the same time.
- Interrupts allow processors to go into low-power modes (sleep, idle, etc.) while waiting for something to happen.
- Polling can give very tightly timed responses to events if there is only one thing for the processor to do.
- Interrupted hardware often takes several cycles to respond to an event.
- Interrupts are preferred when low latency is required.

When an application requires absolute, deterministic timing, polling is preferred. Interruptions, by their very nature, can occur at unpredictable times, which complicates energy usage analysis, whereas polled systems make it simpler to conduct a more structured analysis.

Summary

Logging is a given in modern systems, as it plays a crucial role when problems arise in deployed environments. However, logging can have a significant impact on a system's performance. Similarly, removing looping and making an informed choice on polling techniques appropriate for the application can reduce processing time. All of these factors present opportunities to reduce energy consumption and are techniques that should be considered and used by the software engineer to drive improvements in efficiency.

8. Power adaptation

These are applications and code designed to be adaptive and able to detect the power condition of hardware and adjust processing to maximise power efficiency. The ability for software to adjust power consumption based on the prevailing operating conditions in various real-world environments, particularly mobile applications, remains rare, but using this technique can go a long way towards making a digital product more efficient and sustainable.

Device power optimisation and monitoring

Ahead of using sophisticated next-generation power adaptation solutions and frameworks, different options already exist for optimising power consumption and enhancing power-saving performance. To bring this to life, I have chosen the Android power-saving features as an example of how application design can utilise existing capabilities and adapt power usage to improve device efficiency.

Android has two power-saving features that help extend battery life. These are achieved by managing the behaviour of applications when a device is not connected to a power source. The 'Doze' feature reduces battery consumption by delaying CPU and network activity for applications during extended periods of inactivity. The 'App Standby' feature delays network activity in the background for applications with which the user has not recently interacted. These are good examples of how software engineering techniques can use power adaptation to improve energy efficiency.

Doze

Doze mode is activated when a user leaves a device disconnected with the screen off for a set period of time. In Doze mode, the system restricts installed applications' access to network and CPU-intensive services in order to conserve battery life. Doze will also defer tasks, data synchronisation, and other functions.

As shown in the diagram below, a system can exit Doze for a brief duration to allow applications to complete their delayed tasks. During this 'maintenance' window, the system executes all pending synchronisations, tasks, and alarms and grants network access to applications.

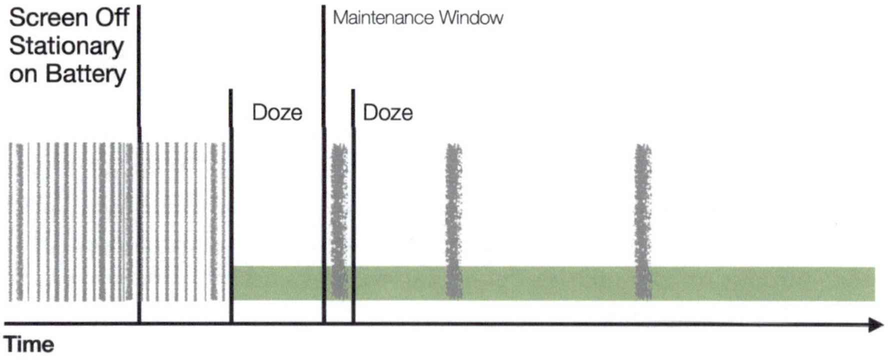

At the end of each maintenance window, the system returns to Doze mode, suspending network access and delaying tasks, synchronisations, and alarms. Over time, the system learns to schedule fewer maintenance intervals, thereby reducing battery consumption during extended periods of inactivity when the device is not connected to power.

The system leaves Doze as soon as the user awakens the device by moving it, turning on the screen, or connecting a charger, and all applications resume normal operation. This is a good example of how an application can be designed for power adaptation based on operating conditions.

Adapting an application to doze

Doze can have different effects on applications, depending on the capabilities they offer and the services they employ. Many applications operate routinely during Doze cycles without modification, while others are optimised to manage networks, alarms, tasks, and data synchronisations.

The Doze restriction on network access can impact an application if it relies on real-time messages such as tickles or notifications. If the application requires a persistent network connection to receive messages, solutions such as Firebase Cloud Messaging (FCM) can be used. I'll cover this more on the next page.

App Standby

App Standby enables an application to determine when a user is not actively interacting with it. The decision is made when none of the following conditions apply:

- The user launches the application.
- The application has a process in the foreground, either as an activity or foreground service or in use by another activity or foreground service.
- The application generates a notification that users see on the lock screen or in the notification tray.

- The application is an active device admin app (for example, a device policy controller). Although they generally run in the background, device administration applications never enter App Standby because they must remain available to receive policy instructions from a server at any time.

A system can release applications from their standby state when the user plugs the device into a power source; this allows them to freely access the network and execute pending tasks and synchronisation requests. Normally, a system permits idle application network access approximately once per day if the device is idle for extended periods.

Other power adaptation considerations

In addition to Doze and App Standby, there are other power adaptation techniques you can consider for your application; these include:

Use Firebase Cloud Messaging (FCM) in idle time

Firebase Cloud Messaging (FCM) is a cloud-to-device service that enables real-time messaging between backend services and Android applications (Apple iOS has a similar service called APNS, or Apple Push Notification Service). FCM provides a singular, persistent connection to the cloud, which can be shared by all applications requiring real-time messaging, per the diagram on the next page[84].

This shared connection optimises battery life by eliminating the need for multiple applications to maintain their own distinct persistent connections, which can rapidly deplete a battery. If an application requires messaging integration with a backend service, FCM rather than a persistent network connection can be used. Similarly, if an application requires downstream messaging, FCM (or a comparable real-time messaging service) is also suitable. Where an application already employs FCM, ensure that it only

---

[84] firebase.google.com, 'FCM Architectural Overview'

sends high-priority notifications when it is absolutely necessary to notify a user.

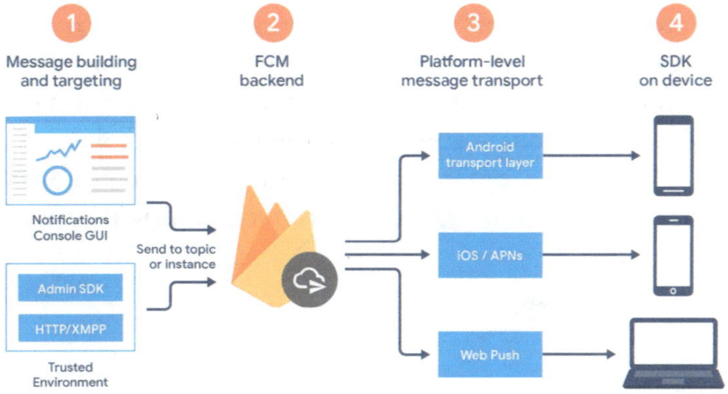

Monitoring power levels and charging states

The impact of application updates on battery life is dependent on the battery level and charging status of the device. Performing updates while the device is charging over AC power is negligible, so it is possible to maximise refresh rates whenever the device is being charged. In contrast, if the device's battery is discharging, reducing the update rate helps extend battery life.

Monitor changes in power level

Continuously monitoring the battery condition is not always necessary or efficient and can have a greater impact on the battery than an application's normal behaviour; therefore, it is recommended to only monitor for significant changes, like when the device reaches a low battery threshold.

Background optimisation techniques

Background processes can be demanding on memory and battery life. For example, a system broadcast might initiate a large number of background processes that have registered to listen to it. Responding to these requests can affect both device performance and the user experience.

Summary

Performance and user experience can be enhanced by optimising applications to operate on low-memory devices or in low-memory conditions. Eliminating dependencies on background services and registered system broadcasts can improve the performance of an application. All applications, especially mobile and IoT software relying on battery power, should be optimised to operate without these background processes whenever possible, as this will reduce energy usage and carbon emissions.

9. Computational accuracy

In some instances, applications using data may not need to produce or calculate absolutely accurate results; acceptable accuracy may be sufficient. A system with a reduced, but controlled, level of computational accuracy can also be more resilient as it is more tolerant to variations in data quality. By limiting the computational accuracy of an application to a level commensurate with the operational requirements at any moment in time, it is possible to substantially reduce compute and network demand, and by doing so, reduce the amount of energy being consumed.

Accuracy, precision, and tolerance

Determining the appropriate level of computational accuracy can be challenging, and applying principles and methods employed in the manufacturing and engineering industries can help. Here, during the design phase, standards for a product's accuracy, precision, and tolerance are established. These fundamental elements of product design are equally applicable in the world of software engineering and code design, and they can be used to set computational accuracy limits. Let's go through each of these elements in turn:

Accuracy

The definition of accuracy is the quantity of measurement uncertainty relative to an absolute standard. In software engineering, accuracy will include the effects of errors caused by gain and offset parameters in your code.

Precision

Precision is the ability to reproduce a measurement. For example, when repeatedly measuring a steady-state signal, if the values are close together, the precision or repeatability is high. The values need not be true values; they can simply be grouped together. The accuracy is the difference between the average of the measurements and the true value.

When discussing random and systematic errors, the terms 'accuracy' and 'precision' are frequently used, but they should be applied with care. As stated previously, 'accuracy' refers to the proximity of a measurement to its true value. On the other hand, 'precision' is a description of repeatability and is sometimes used incorrectly to denote accuracy. The diagram below illustrates the differences between accuracy and precision when attempting to hit the centre of a target.

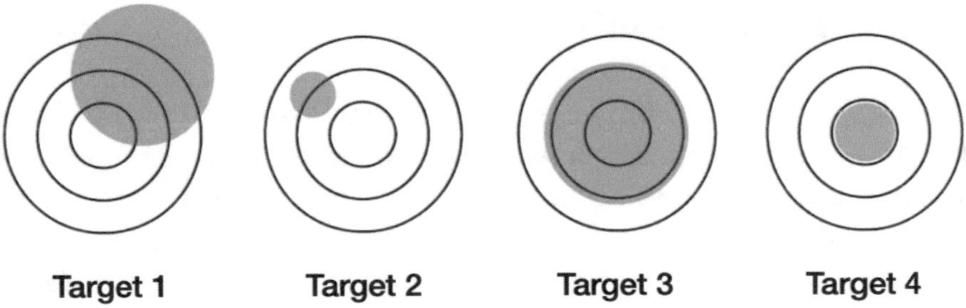

Target (1) shows inaccurate and low-precision hits on the target.

Target (2) shows better precision but a similar level of inaccuracy. Both Target (1) and Target (2) show a consistent offset, or bias.

Target (3) is more accurate.

Target (4) is the most accurate and precise.

Sometimes, 'precision' is not regarded as a useful measurement term, and 'accuracy,' especially in statistical analysis, is more commonly described as uncertainty. For the purposes of understanding computational accuracy, it is important to be familiar with how both these terms and definitions are used.

Tolerance

Tolerance is the permissible range of variation for a value. In software terms, tolerance defines the permissible limits of variance required for an application to meet user or business needs.

Tolerance counterbalances extreme precision and accuracy, as it helps to identify an acceptable amount of deviation or variation. An acceptable tolerance value depends on the application and the end-user or system requirements it is intended to satisfy and is a crucial factor when determining the level of computational accuracy for your application.

Statistical approaches

When determining computational accuracy, statistical methods and techniques are needed to provide the foundation for organising and simplifying data and allowing objective estimation to demonstrate that an analysis is under control or explain why a change has occurred. Using statistical methods, it is possible to make sense of large amounts of data and help determine the application's tolerance limits and computational accuracy.

Common approaches and methods include:

Mean

The mean is the average of a set of values. The most common method of calculation is a simple arithmetic mean, which involves adding up the numbers and dividing them by the total number of observations.

Standard deviation

The standard deviation is a measure of a set of values' variation or dispersion. A low standard deviation indicates that the values tend to cluster near the set's mean, whereas a high standard deviation indicates that the values are dispersed over a larger range.

Regression

Regression is a statistical method for analysing and understanding the relationship between two or more relevant variables of interest. Adapting a procedure to perform regression analysis helps to determine which factors are significant, which factors can be ignored, and how they influence one another.

Hypothesis testing

Hypothesis testing is a form of statistical inference that uses data from a sample to derive conclusions about a parameter or the probability distribution across a population. Normally, a tentative assumption about the parameter or distribution is made, followed by the formulation of an alternative hypothesis. Using sample data, the hypothesis-testing procedure determines whether the tentative hypothesis can be rejected or whether the alternative hypothesis is true.

Sample size determination

Sample Size Determination involves selecting the number of observations or replications for a statistical sample. The sample size is an important feature of any empirical study in which the aim is to make inferences about a population from a sample.

Confusion matrix

A Confusion matrix is a common method of solving classification problems. It can be applied to binary classification problems as well as problems involving multiple classes.

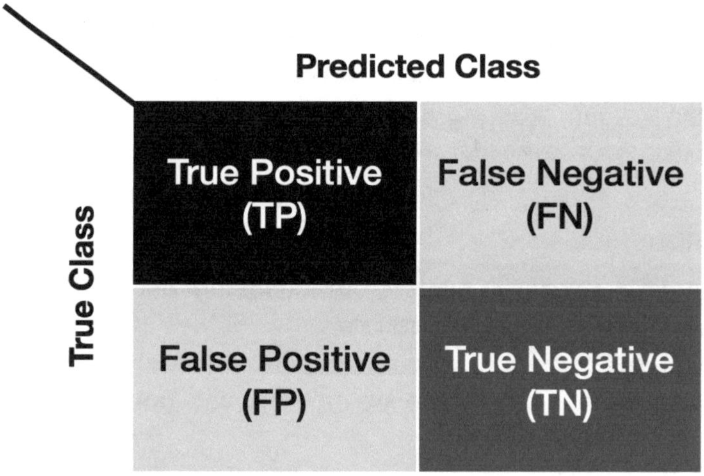

Confusion matrices represent the number of actual and predicted values. In the diagram above, the output 'TN' indicates the number of True Negative examples that were accurately classified as negative. Similarly, 'TP' stands for True Positive and represents the number of positive examples that have been correctly classified. The term 'FP' indicates a False Positive value, which is the number of actual negative examples classified as positive, whereas 'FN' indicates a False Negative value, which is the number of actual positive examples classified as negative. Applying a confusion matrix to a dataset can help to better understand the computational accuracy requirements for a software application.

Other considerations

Error

An 'error' in measurement is the difference between the recorded value or measurement and the true, theoretically correct value. Software engineers take measures to mitigate or prevent the effects of errors, whether intentional or unintentional. During the design phase, projects must be thoroughly specified in order to allow for and cater for errors. This includes identifying the error 'tolerance'.

Uncertainty

Uncertainty is the degree of doubt we have regarding any measurement or calculation. In contrast to tolerance, which is specified by software engineers and expressed as a value, uncertainty is defined statistically.

During the design phases of a project, software engineers must define the uncertainty of their system to ensure that it meets all requirements. The agreed-upon tolerance value has to account for measurement uncertainty and will likely be part of, or inform, the product's acceptance criteria.

Summary

After gaining an understanding of how to calculate computational accuracy, it is possible to consider opportunities to intentionally regress from a baseline position. Using different statistical methods and the basic principles of accuracy, precision, and tolerance, it is possible to find an acceptable level of computational accuracy that meets the needs of either the end user or the system, and ideally both. By evaluating and incorporating these concepts and techniques into software design, it is possible to reduce the computational load of an application, thereby enhancing its performance and reducing its energy consumption.

10. Monitor energy consumption

Monitoring the real-time energy consumption of an application can help identify which modules can be optimised to reduce emissions. All hardware, when powered, will consume energy even when it is idle; therefore, it is necessary to measure the baseline power consumption versus the power consumption when the application is operating to ensure that actions can be taken to reduce power consumption in the appropriate application state. Some measurement tools are able to assess application-level energy consumption using data collected from various system interfaces. Whilst other tools make use of energy models, utilising algorithms created to measure the energy consumption of a device. Using a combination of these techniques, it is possible to profile a system and determine which system functions contribute the most to power consumption.

All of this energy consumption data helps to establish a baseline against which it is possible to design greater energy efficiency and identify carbon reduction opportunities.

Optimising for lower energy consumption

New developments in low-power management have launched a wide range of ultra-low-power sleep modes that provide more acceptable levels of granularity beyond just run, idle modes, standby, doze, sleep, and deep sleep. The goal is to programme an application to spend as little time as possible in the 'active' mode of the Microcontroller Unit (MCU). It can involve simplifying calculations, clustering operations, or switching to an asynchronous, interrupt-driven design.

The following are useful techniques to consider when optimising for lower energy consumption:

- Increase the amount of time a device spends in a low-power sleep state.

- Increase clock speed to finish processing faster and use low-power timers for wake-up.
- Efficient coding is needed to ensure applications are not performing unnecessary tasks.
- Consider the effects of firmware choices and updates.
- Avoid excessive push notifications.
- Use Smart Monitoring and predictive maintenance techniques.

As an example, consider an application in which the MCU is clocked at 16 MHz and is woken up every second to execute a 700-instruction-long task. One of the design requirements might be that the device can operate for at least five years on a single battery. If it were possible to eliminate just one instruction (assuming one clock cycle) from this code, the energy savings would enable the system to be powered for an additional 2.01 days over the product's five-year lifetime. Although one instruction and 2.01 days over five years may not sound significant, it demonstrates how code changes can force more efficient use of the MCU.

To optimise code for increased efficiency and performance, consider the following techniques:

Short Variables

Even efficient compilers frequently overlook a number of techniques and methods for optimising code for low power consumption. On some 32-bit MCU architectures, such as ARM, it is impractical to deal with single-byte variables due to 32-bit registers and sufficient memory being readily available.

Avoid using 'small or short' variables, for instance, if there is lots of memory. This will not save energy, as the entire memory array will still need to be powered. Although it will save one byte of data memory, it will require two extra instructions, per the code example below.

```
;unsigned short int counter = 10;
;while (counter)
;{
 MOVS r0,#0x0A
loop:
;// do something
; counter- -;
;}
 SUBS r0,0,#1
 LSLS r0,r0,#16
 LSRS r0,r0,#16
 BNE  loop
```

→

```
;unsigned short int counter = 10;
;while (counter)
;{
 MOVS r0,#0x0A
loop:
;// do something
; counter- -;
;}
 SUBS r0,0,#1
 BNE  loop
```

Adjust to clock cycles

A module operator checks the remainder of an integer division operation, and in most MCU instruction sets, the execution of the division instruction requires more than one clock cycle if the results need to be immediately accessible. Changing to more but simpler instructions (i.e., a single set of instructions) actually conserves both time and energy, per the code example below.

```
if ((I%5) = = 0)
{
// do something
}
i++;
```

→

```
if (counter = = 5)
{
// do something
 counter = 0
}
else {
 counter++;
}
i++;
```

In-lining

In-lining is a popular technique for performance optimisation. In in-line functions, rather than generating code to execute the function, the function's body is inserted wherever the function is called. In the example below, with three functions (Function A, B, and C), Function A is executed and calls Function B from within its body, and Function B calls Function C. Within each function, returns are unavoidable. At least two calls and two returns will be eliminated through in-lining, resulting in an increase in code size but a decrease in execution time. The majority of compilers will automatically perform in-lining, but only if the feature is enabled.

```
        function_A(...)
:
CALL        function_B(...)

CALL                function_C(...)

RET
RET
```

→

```
function_A(...)

function_B(...)

function_C(...)
```

Loop-unrolling

Loop-unrolling is a widely applied technique. The primary purpose of loop-unrolling is to reduce the number of conditional branches. In the following example, we want to copy the values of a 64-byte array. If the body is duplicated eight times, the size of the code will increase, but fifty-six conditional jumps will be eliminated because the loop will only be executed eight times. Examine the final code the compiler produces because it typically only applies a small amount of unrolling. The most efficient method to improve execution speed would be to unroll the whole loop in this case.

```
unsigned int i =
64;
while ( i  != 0)
```

→

```
unsigned int i = 64;
while ( i  != 0)
{
  A [ i ] = B [ i ];  i - - ;
```

Code calculations

In addition to the variable types, instruction usage, and code flow, the arithmetic and calculations used within the code can also be optimised. The following example is a simple form of code calculation optimisation. First, add the variables a, b, and c and store the result in variable x (two ADD instructions are required). The sum of variables a, b, and d is then stored in variable y (this requires two additional ADD instructions). It is possible to use the sum of a and b and add the variables c (in the first case) and d (in the second case) to this intermediate result. This will save a minimum of one ADD instruction. Typically, compilers do this automatically, and advanced compilers can optimise significantly more complex mathematical expressions and equations. In very complex cases, it may be necessary to call a mathematical function from a standard library.

```
X = a + b + c              temp = a + b
y = a + b + d      ➔       x = temp + c
                           y = temp + d
```

Standard library

Use of a standard library mathematical function will be required in more complex scenarios, for example, calculating the geometric mean for a single precision floating point. To calculate the geometric mean of six values, using optimised mathematics, the sixth root could be parsed as the square root of the cube root of the argument. The geometric mean of six values can be calculated as follows:

$$A = \sqrt[6]{x_1 * x_2 * x_3 * x_4 * x_5 * x_6} \quad \Rightarrow \quad A = \left[\left(x_1 * x_2 * x_3 * x_4 * x_5 * x_6 \right)^{\frac{1}{3}} \right]^{\frac{1}{2}}$$

Always examine the available standard library functions. For example, use the standard function 'pow' (power) to calculate any power of the argument, but sometimes cube and square root functions are more convenient than the standard 'pow' function, which uses a universal raw algorithm to calculate any power of the number that is provided. Cube and square root algorithms are better optimised. The final result might be as follows:

```
float a, b, c, d, e, f, result;
... initialisation of variables

temp = a * b * c * d * e * f;
result = powf (temp,  (1.0f / 6.0f) ) ;
```

➔

```
float a, b, c, d, e, f, result;
... initialisation of variables

temp = a * b * c * d * e * f;
result = cbrtf (temp) ;
result = sqrtf (result) ;
```

Here, the 'powf' function takes three times longer to execute than the optimised cube root and square root functions. When speed optimisations are enabled, and in conjunction with a modern, high-performance compiler, this can result in highly efficient code.

Optimisation of peripherals

In advanced microcontrollers, in addition to the core, there can be additional bus masters. Direct memory access (DMA) controllers are among the most powerful co-workers, as they can be used to automatically perform simple tasks that would otherwise require polling or interrupts. Again, let's use a simple example to illustrate how this might work in practice. Here, an analogue-to-digital converter (ADC) is used to measure the signal on an analogue input port, after which the sampled values are processed and the results are transmitted using the USART/SPI communication protocol (Universal Synchronous/Asynchronous Receiver Transmitter, which is hardware enabling a device to communicate using serial protocols; and Serial Performance Interface, which facilitates short-distance communication between a MCU and peripheral circuits).

The steps would be:

Step one – sampling

- ADC and DMA is ON.
- ADC is sampling the preconfigured analog input channel.
- DMA is storing the converted values into the embedded SRAM memory.
- (CORE is kept SLEEPING during this period)

Step two – processing

- ADC and DMA is put OFF.
- CORE is woken up and the sampled data are processed (average value, calculation, geometric mean, or any other filtering).
- (CORE is RUNNING during this period)

Step three – sending

- USART/SPI and DMA is put ON.
- Processed data are transmitted over USART/SPI.
- DMA is providing the data stored in SRAM (Static Random Access Memory) to the communication peripheral.
- (CORE is kept SLEEPING during this period)

During the operation of the DMA, the core can perform other duties or be shut down, resulting in a significant reduction in average power consumption and total energy consumption.

Summary

Take advantage of all available optimisation-based efficiencies, but be mindful of the architecture and compiler's limitations when working at the code level. It is possible to use these techniques to extract maximum energy savings from the architecture, peripherals, and code in each clock cycle, as well as from the compiler, in order to achieve the lowest power consumption and highest performance.

11. Programming language efficiency

Your choice of programming language will have an impact on your software product's energy efficiency. According to the Online Historical Encyclopaedia of Programming Languages, there are around 9,000 coding languages. Wikipedia claims that 700 of these are commonly used[85]. But which ones are better, and why? This section will explore the answers to this question and examine the results of academic research that has been undertaken on the topic.

There are a number of factors that affect a programming language's speed and efficiency. The main areas to consider are:

Choice of compiler

A compiled language is more likely to be faster than an interpreted language because it compiles the entire programme into executable machine code at once, whereas an interpreter converts the programme line by line into machine code. A compiler converts your selected high-level language into intermediary code, which saves time during execution, and as a result, the computer executes the intermediary code more quickly than it would with an interpreter.

Cross-platform usage

In certain object-oriented languages, such as Java and C#, the source code is both compiled and interpreted. Due to cross-platform usage and conversion into byte code, the performance of the cross-platform language will be compromised, making it slower than some other languages.

Security

Some programming languages send source code to multiple layers of security for checking, which slows down the compilation process and

[85] Source: hopl.info and wikipedia.org, 'List of Programming Languages'

increases execution time. Take time to investigate this for your code, as it can add processing time and energy consumption.

Statically typed vs. dynamically typed

In statically typed programming languages (e.g., Java, C, C++, and C#), checking takes place at compile time, where the source code in the selected programming language is converted to machine-readable format. This means that when a variable is statically typed, the type associated with each and every variable must be known. The challenge with statically typed languages is that the detection of errors in variable-type associations will stop the compilation process until the errors have been fixed.

Dynamically typed means that the variable type is checked at runtime, and any errors are reported at runtime. Examples of dynamically typed languages include Python, Javascript, Perl, and Ruby. Dynamically typed languages offer more flexibility but with less optimised code, while statically typed languages offer more optimised code with less flexibility.

Analysis and benchmarking

The Computer Language Benchmarks Game (CLBG)[86] is a well-known set of programming language benchmarks used to generate reproducible test scenarios.

Using a standard CLBG setup (including Intel's Running Average Power Limit (RAPL) tool, alongside Ubuntu desktops running 3.20GHz Intel Core i5-4460 Haswell processors with 16GB of RAM, with Ubuntu Server 16.10 with Linux 4.8.0-22), academic and research-based tests have been conducted on runtime, memory utilisation, and pure energy consumption. In these experiments, paradigm alignment was undertaken per the table below,

---

86 Source: https://benchmarksgame-team.pages.debian.net/benchmarksgame/

where programming languages with similar purposes were comparatively evaluated.

Paradigm	Languages
Functional	Erlang, F#, Haskell, Lisp, Ocaml, Perl, Racket, Ruby, Rust
Imperative	Ada, C, C++, F#, Fortran, Go, Ocaml, Pascal, Rust
Object-Oriented	Ada, C++, C#, Chapel, Dart, F#, Java, JavaScript, Ocaml, Perl, PHP, Python, Racket, Rust, Smalltalk, Swfit, TypeScript
Scripting	Dart, Hack, JavaScript, Jruby, Lua, Perl, PHP, Python, Ruby, TypeScript

The CLBG testing demonstrated that compiled languages are more efficient. The table below lists the performance in rank order in each column.

<table>
<tr><th>Time & Memory</th><th>Energy & Time</th><th>Energy & Memory</th><th>Energy & Time & Memory</th></tr>
<tr>
<td>C | Pascal | Go
Rust | C++ | Fortran
Ada
Java | Chapel | Lisp | Ocaml
Haskell | C#
Swift | PHP
F# | Racket | Hack | Python
Javascript | Ruby
Dart | TypeScript | Erlang
JRuby | Perl
Lua</td>
<td>C
Rust
C++
Ada
Java
Pascal | Chapel
Lisp | Ocaml | Go
Fortran | Haskell | C#
Swift
Dart | F#
JavaScript
Racket
TypeScript | Hack
PHP
Erlang
Lua | JRuby
Ruby</td>
<td>C | Pascal
Rust | C++ | Fortran | Go
Ada
Java | Chapel | Lisp
Ocaml | Swift | Haskell
C# | PHP
Dart | F# | Racket | Hack | Python
JavaScript | Ruby
TypeScript
Erlang | Lua | Perl
Jruby</td>
<td>C | Pascal | Go
Rust | C++ | Fortran
Ada
Java | Chapel | Lisp | Ocaml
Swift | Haskell | C#
Dart | F# | Racket | Hack | PHP
JavaScript | Ruby | Python
TypeScript | Erlang
Lua | Jruby | Perl</td>
</tr>
</table>

The results for energy, time, and memory were normalised per the table on the next page, clearly showing the performance attributes for different programming languages:

Energy		Time		Memory	
C	1.00	C	1.00	Pascal	1.00
Rust	1.03	Rust	1.04	Go	1.05
C++	1.34	C++	1.56	C	1.17
Ada	1.70	Ada	1.85	Fortran	1.24
Java	1.98	Java	1.89	C++	1.34
Pascal	2.14	Chapel	2.14	Ada	1.47
Chapel	2.18	Go	2.83	Rust	1.54
Lisp	2.27	Pascal	3.02	Lisp	1.92
Ocaml	2.40	Ocaml	3.09	Haskell	2.45
Fortran	2.52	C#	3.14	PHP	2.57
Swift	2.79	Lisp	3.40	Swift	2.71
Haskell	3.10	Haskell	3.55	Python	2.80
C#	3.14	Swift	4.20	Ocaml	2.82
Go	3.23	Fortran	4.20	C#	2.85
Dart	3.83	F#	6.30	Hack	3.34
F#	4.13	JavaScript	6.52	Racket	3.52
JavaScript	4.45	Dart	6.67	Ruby	3.97
Rocket	7.91	Racket	11.27	Chapel	4.00
TypeScript	21.50	Hack	26.99	F#	4.25
Hack	24.02	PHP	27.64	JavaScript	4.59
PHP	29.30	Erlang	36.71	TypeScript	4.69
Erlang	42.23	JRuby	43.44	Java	6.01
Lua	45.98	TypeScript	46.20	Perl	6.62
JRuby	46.54	Ruby	59.34	Lua	6.72
Ruby	69.91	Perl	65.79	Erlang	7.20
Python	75.88	Python	71.90	Dart	8.64
Perl	79.58	Lua	82.91	JRuby	19.84

Summary

Academic research and the CLBG tests have shown that faster languages consume less energy. Therefore, the choice of programming language is crucial to the overall energy efficiency of the software product or application being developed, and great care must be taken to select the appropriate language for the application's specific needs.

In situations where a programming language can't be changed for a more efficient one, consider if it is possible to compile the machine code ahead of deployment. In these scenarios, avoid translation by an interpreter while the application is running, as some compilers for certain interpreted languages, such as PHP, generate interpreted code that requires more processing power, resulting in more energy being consumed to perform the same task. In other cases, small modifications to the application's source code can make it suitable for compilation, allowing it to be run with enhanced performance and reduced energy consumption.

12. Reuse

Software reuse has a wide range of advantages and benefits, such as increasing reliability, accelerating the pace of development, increasing the productivity of software, minimising operational and maintenance costs, enhancing system interoperability, developing software with less manpower, producing standardised software, and ultimately assisting in the delivery of higher-quality software.

Reuse in software can be compared to recycling in the physical world. The more you can reuse an existing asset, whether it's applications, components, or code, the more efficiencies you gain by not duplicating previously expended effort. By increasing the number of reusable assets, you can reduce energy consumption during the production of a software product and, as a result, its carbon footprint. It should go without saying: try to only reuse code that you know is energy efficient.

Maximise code reuse

Reusing code is a common technique for increasing productivity, particularly on projects that utilise open source frameworks. Typically, the concept of code reuse is employed in two main ways:

- Reusing previously written code that was developed internally
- Reusing external code to fulfil a project's requirements

To meet the rising demand for digital products and services, it is important to ensure that any code considered for reuse is safe and of high quality, and that any libraries to be reused are secure and taken from a reliable source.

Reuse can be implemented across multiple areas, including application development, embedded systems, and web development, and is applicable across many programming languages. It facilitates rapid application development at low cost and encompasses the reuse of various assets, such as applications, components, objects, and functions. Code reuse has now become a mainstream technique in most modern development environments.

The following table highlights some common scenarios and techniques for implementing code reuse in support of rapid application development. Adopting the right implementation approach suited to the right scenario can significantly improve your chances of success and achieve the most benefit from code reuse.

Code Reuse Scenarios	Implementation Approach
Internal	The organisation reuses its own developed components internally. Often this involves copying existing code into new build and development projects. Special attention should be placed on code redundancy in these internal use case scenarios.
External	Using third party licensed software components and integrating them into your project. Ensure upfront time is allocated to select the right products or components to reuse according to your requirements and specification.
Planned	Use of tactically designed building blocks of code that have been created to meet the future needs of your oganisation.
Opportunistic	Early stage project identification of existing software and code components that can be redeployed to meet the specific factors and outcomes of the product being delivered.
Referenced	Identifying referenced code components of a specific version that meets the form and structure of the project ocde base.

Ways to reuse code

There are a range of techniques and approaches to consider when creating code that can be reused; these include:

Modularisation

Modularisation is a method for producing code of high quality and versatility and can be used to divide a software system into standalone modules, each of which performs a specific function. This makes the code more manageable and can help eliminate the need for repetitive code writing.

Cohesion

Cohesive code can support reusability across multiple projects and is capable of adapting to various environments by forming relationships between different modules of a class or system.

Loose Coupling

Often, the code used by software modules is portable and unaffected by external linkage; this is when software engineers must manage the connections between functions and classes to avoid zero coupling, as this is not an optimal state. Achieving the proper design balance where the code is loosely coupled can help improve performance.

Utility Classes

Using class methods for basic data formatting and sharing functions, constraint validation, image sharing, compression, and image downloads should be considered best practice and can maximise the reuse of digital assets and code.

Code Flexibility

Code flexibility helps to create code that can be reused in future projects with minimal modification and meets its intended purpose without error.

Decoupling

Ensure that each module performs a single function. A method for file compression, for instance, must only accept a file as input and return the compressed file as output. This technique will decouple file compression as a function from the rest of the code base so that it can be efficiently reused elsewhere in a project.

Inheritance

Object Orient Programming (OOP) paradigms use inheritance techniques extensively. Here, developers can easily repurpose functions to override and modify the behaviour of various methods for multiple objects in the code base.

Libraries

The use of software libraries and Dynamic Linked Libraries (DLLs) is now commonplace. A number of third-party libraries have been developed specifically to perform unique functions and facilitate code reuse. Also consider that a library can contain code for several functions and can be accessed via an API, further encouraging reuse.

Forking

Forking is a basic Git operation that entails modifying someone else's code to meet the requirements of a different project and then distributing it under a different tag. Software engineers who are collaboratively developing source code use Git. This exercise is useful for open source initiatives, such as modifying existing frameworks and creating new ones.

Methods and function calls

Function calls are modifications made in a single location that are then reflected everywhere the function is called. This permits a single code modification to satisfy the needs of multiple users and can be highly efficient.

Don't Code

Simply, don't write lines of code that are not needed.

Code reuse in practice

One of the primary advantages of code reuse is the production of a higher-quality final product within a reduced development time. It is important to reuse code from trusted sources and to resolve any vulnerabilities prior to deployment in order to improve or maintain operational performance. It is advised to keep records of every third-party component in use so that assets can be readily tracked and reported. Consider the following methods to manage code reuse in practice:

Git submodules

A Git submodule enables the creation of a subdirectory for another Git repository. Submodules can contain shared code for use in multiple applications. However, because of coupling between projects, it can be more difficult to administer a cross-repository project because submodules may not always receive updates from their parent directory. If you can take steps to avoid this, then Git submodules will magnify the reuse opportunities for your code.

RAD stack

Utilising a modern, open-source stack is a common method to simultaneously support all platforms and reuse frontend components. As an open source framework, the RAD stack (React Native and Django) includes a large number of open source libraries and cross-site scripting to protect an application from vulnerabilities. There are many RAD-scalable APIs that can provide users with access to third-party services as well as the option to add and remove services.

Monorepos

A monorepo is a version-controlled code repository holding many projects and sub-projects. Monorepos allow multiple packages to be stored, enabling components to be configured, unified, and published. Its drawbacks include extensive refactoring and dependency graph management. Having a large repository holding many commits, branches, files, and content can aid reuse but can also lead to performance issues, so a careful balance between size and performance needs to be achieved to get the best from a monorepo.

Bit

This popular and in-demand JavaScript tool facilitates code reuse by controlling unified code modifications through a component dependency tree. Its workflow defines and monitors shared components in all projects automatically, displays their dependency graphs, and does not require refactoring for reuse. Components can be shared on a remote server where they can be deployed as npm packages (npm is the world's largest software registry with over 800,000 code packages, and it's where open-source developers share their software). If modifications or updates are made to a single project, those modifications can be synchronised with other projects by updating the component's version.

Summary

Current application development standards support the creation of reusable building blocks of code, designed as reusable segments, distributed across a development community, and are open-sourced so that developers can adopt, modify, and collaborate effectively. In addition to the traditional advantages of code reuse, such as faster delivery and enhanced resilience, there are opportunities to reduce energy consumption during the design and development phases of an application. By minimising time and effort through reuse and incorporating the techniques into an overall reuse strategy for a digital product, carbon costs can be reduced.

Conclusion to Part Two

Sustainable software engineering is a growing trend that seeks to design and build software that improves application efficiency, reduces energy consumption, and minimises the long-term impact of IT infrastructure on the environment. Increasing demand for carbon-free and low-energy software has made energy efficiency a new standard for the modernisation of digital products and services.

If we adopt the same software development considerations as more than two decades ago, when coding was restricted to strict lengths, sizes, and compute capacity, coding can become more environmentally friendly almost immediately. However, the availability of inexpensive, nearly limitless computing resources has resulted in 'bloatware' and energy-inefficient software.

Achieving a carbon-neutral or Net-Zero digital future requires a focus on the energy consumption of software. Using techniques that limit energy consumption, such as reducing the file size of text, images, or video, and media compression tools can reduce data sizes significantly, thereby enhancing the user experience, accelerating navigation, and reducing the strain on the technology value chain. Developers of websites must understand when to use high-quality multimedia and when smaller files can achieve the same purposes.

The Sustainable Software Engineer is needed to develop the next generation of energy- and carbon-efficient digital solutions. Utilising sustainable software engineering principles and the twelve core engineering techniques that support these principles is crucial for bringing about the changes needed. These principles and techniques should be included in all software development toolkits to enable the development and operation of greener, more environmentally responsible software.

Practical - Sustainable Software Engineering Checklist

As an aid and reminder of sustainable software engineering principles and techniques, have a look at the checklist on the next page. Consider using it on your next project or phase of code development to challenge your engineering choices and start taking steps to build sustainability into your software product.

You can also access and download the checklist using the QR code above, or alternatively, visit: www.gocode.green/sustainabilityassessmenttools

GoCodeGreen
Decarbonise Digital®

Sustainable Software Engineering Checklist

Adopting Green Coding Principles		Tick all that apply
Engineering Energy Efficiency	Ensured your application is taking responsibility for the electricity it consumes and considered whether you have designed it to consume as little energy as possible	☐
Carbon Efficiency	Minimised the amount of carbon emitted per unit of work and wherever possible, made choices to consume electricity with the lowest carbon intensity available	☐
Hardware Efficiency	Focused on extending the lifespan of hardware through more efficient use of resources (e.g., running your workload on as few servers as possible with high utilisation rates) and ensured that your application has been designed to be able to run on older hardware components	☐
Data Efficiency	Made choices to reduce the amount of data and distance it has to travel to reduce energy demand and associated carbon emissions	☐
Measurement	Selected tools that can help measure, baseline and identify actions to reduce the carbon impact of your software, and used the data to set targets that you can track and to monitor performance	☐

Applying Sustainable Software Engineering Techniques		Tick all that apply
Software Power Consumption	Considered using appropriate energy profiling tools and selected an energy consumption technique suitable for your code / project	☐
Data Caching	Selected a data caching technique with appropriate attention on approach and configuration to help improve energy efficiency, and have actively considered whether data caching on server or remotely via a network is most suitable for application	☐
Data Exchanges	Selected an appropriate data exchange method to more efficiently process, transfer and store data	☐
Data Compression, Aggregation and Retention	Applied effective data compression methods, data aggregation techniques appropriate to your application, and selected a data retention policy that balances the processing and data storage requirements of your software	☐
Image Sizing	Adopted various optimisation techniques and format choices to reduce image size and improve the energy efficiency of your website or image intensive application	☐
Unused Features	Assessed and removed unused features and code to improve energy efficiency and increase carbon savings	☐
Logging, Loops and Polling	Devised an efficient and optimised logging strategy and identified and removed loops from your code, and have considered how polling affects performance and implemented changes in your application	☐
Power Adaptation	Applied techniques to monitor power levels and optimise power consumption based on available power features relevant to your application	☐
Computational Accuracy	Applied accuracy, precision and tolerance based methods to your application and code, and made choices that will lead to reduced data processing transfer and storage to help lower energy consumption	☐
Monitoring Energy Consumption	Considered techniques to optimise your code and application for lower energy consumption, balanced performance with energy usage and devised approaches and selected tools to monitor energy consumption	☐
Programming Language Efficiency	Reviewed and considered your programming language, compilation and interpretation choices and balanced selection based on the different dimensions of energy, time and memory usage	☐
Reusable Assets	Devised a code reuse strategy and selected balanced methods and approaches based on deployment challenges whilst lowering energy consumption	☐

Part Three: Green Design

An Introduction to Green Design

Apart from engineering software more sustainably, what more can we do to achieve a low-carbon digital future? I believe that how we approach the development of all future software products will determine the extent to which we succeed, and the way in which we approach the design of digital products and services will be critical to that success.

The following quote from the EU Science Hub emphasises the importance of good design and how it can have a significant influence on the sustainability and environmental impact of the software solutions we develop:

> **"It is estimated that over 80 percent of all product-related environmental impacts are determined during the design phase of a product."**
>
> Source: EU Science Hub

In this part of the book, I will review the fundamentals of green design-thinking before moving on to discuss green design principles, and then reinforce this by talking through real-world examples of green design techniques. This combination of thought processes, principles, and techniques can be grouped together and called 'green design'.

Before even writing a single line of code, we can ensure that we are addressing the right problems, and this is where green design-thinking is critical. To achieve this, we need to incorporate sustainability into the core of all future problem-solving and solution-development.

Design really matters

If we had used green design to build our current digital landscape, it wouldn't be the major cause of carbon emissions that it is today. It's still not too late to play our part. We can use green design alongside sustainable software engineering practices to deliver the lower-carbon digital products and services of the future and make our contribution to averting climate change.

Regardless of your area of design specialisation, whether it be Service, UX, Software, Sustainable, Circular, Responsible, Web or Design to Value, it is essential that your approach is built on a solid foundation based on design methods, thought processes, and structure. I believe that 'design-based thinking' should serve as the basis for this foundation, and over the next few sections, I'll explore how we can make this thinking 'greener'.

> "Thinking about design is hard, but not thinking about it can be disastrous"
>
> *Ralph Caplan*

The challenge of introducing and embedding 'green' into design-based solutions can be made simpler by using design-based thinking at the start of any new project. This would enable designers, architects, and software engineers to carefully consider the decisions they make, the prototypes they develop, and the solutions they deliver alongside the user, customer, and service context of the product.

I cannot overstate the significance and potential impact of design stage consideration and problem solving in reducing the environmental impact of any new digital product. You are creating the digital solutions and landscape of the future, so wouldn't it be wonderful if you could balance the amazing features and customer experiences you create with minimising any negative environmental impacts related to your digital solution? This part of the book will explore how we can achieve that.

Practical - Green Design Self Assessment

If you are a designer or have a design background, complete the Green Design Self Assessment. This will provide you with a rating based on your knowledge and of the principles and techniques described in this section of the book. The ratings are derived from insights gained from GoCodeGreen's assessments across multiple industries and organisations.

You can access the self assessment using the QR code above, or alternatively, visit: www.gocode.green/sustainabilityassessmenttools

Design Thinking

Before diving into the specifics of green design principles and techniques, let's establish a context that enables an in-depth consideration of the various problems, challenges, and opportunities that will arise as any new digital solution begins to take shape. This will form the context for considering how sustainability can become a core element in all future design thinking.

The Double Diamond model[87], which was initially developed by the Design Council in the United Kingdom (and different from 'Design Thinking', a process standardised in the 1990s by professors from the Stanford Design School who founded IDEO, a design consultancy and marketing firm that sold design as a consulting package), is a perfect place to start building this context. I prefer it as a more grassroots approach to design-based thinking and as a means for incorporating sustainability into the design process.

Service, product, and UX designers often cite the Double Diamond as their standard, go-to process, and it is common for them to reference the Double Diamond when explaining their approaches. Product managers have also adopted the Double Diamond in the product discovery process to aid in communicating and prioritising problems as opposed to only focusing on solutions. Often, you will hear designers describe their solutions through the Double Diamond lens.

The Double Diamond model can act as the foundation for green design; it is a strong visual model for understanding the design process and a great tool for making sense of design as it focuses on both problems and solutions, per the diagram on the next page.

---

[87] Credits & Acknowledgements:
Throughout this section we will be referring to the Double Diamond Design Thinking approach, all credit and copyright is attributed to the Design Council per Attribution 4.0 International (CC BY 4.0).

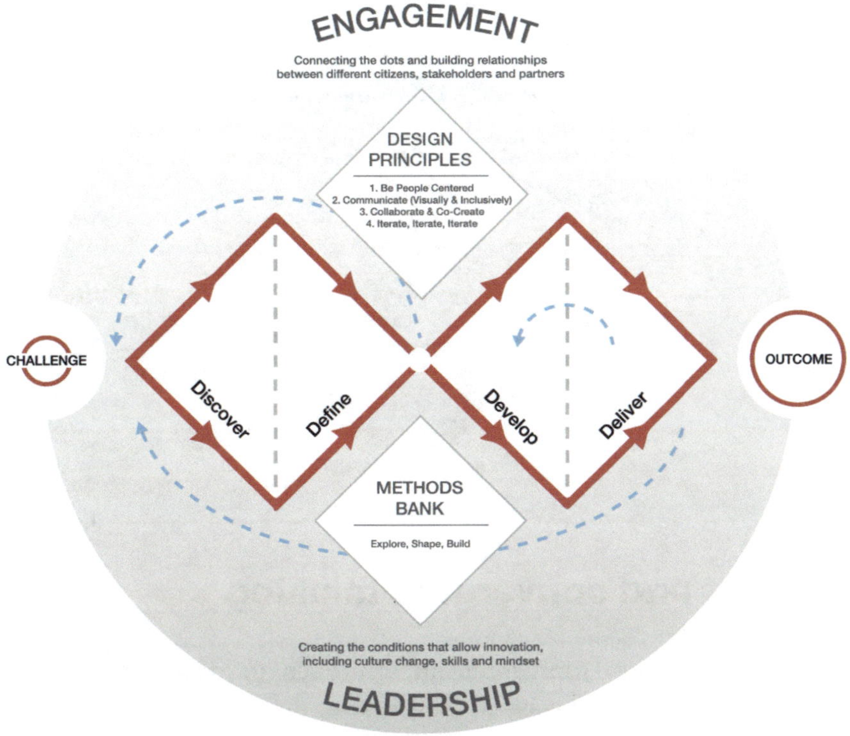

The Double Diamond contributes to the creation of value by emphasising problem analysis before solution development. By incorporating the challenges associated with achieving sustainability into the model, we can examine how design thinking can help accommodate and bring solutions to them.

Double Diamond design principles

Per the Design Council diagram above, it could be argued that incorporating sustainability into the Double Diamond could fall under 'Design Principle 1: Be People-Centred', but for the purposes of explicitly incorporating the 'green' element into design thinking, let's add a fifth design principle to test the model, as indicated by the green arrow.

Divergent and convergent thinking

The Double Diamond represents an approach to design that combines divergent and convergent ways of thinking per the diagram below.

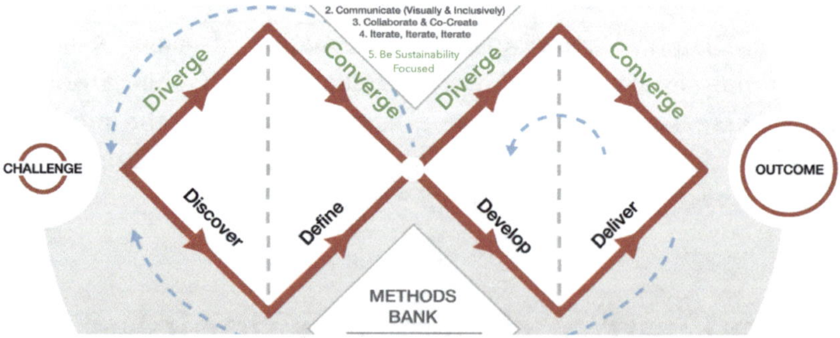

Divergent thinking is free-flowing, non-linear, and considered to be 'creative'. It helps in exploring multiple perspectives and angles. Convergent thinking is focused and linear and helps narrow down to the 'right' answer. It helps refine the definition of the project's scope and can be viewed as a counterbalance to divergent thinking. These two methods of thinking form the diamond shapes of the model.

Problem space and solution space

In the Double Diamond model, problem-finding and problem-solving, two separate activities in the design-based thinking procedure, represent the problem and the solution as being equally important.

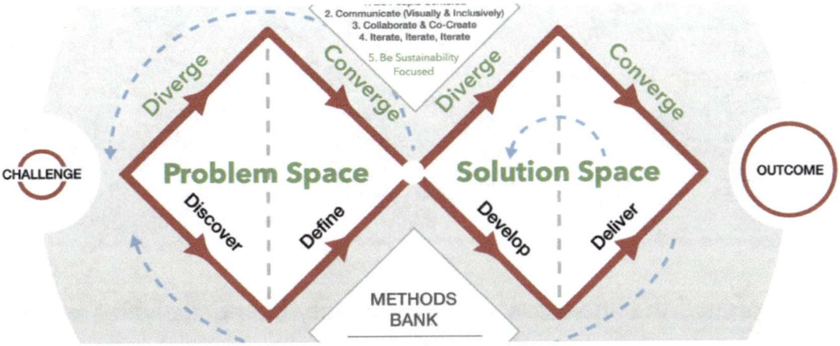

In the diagram above, the first diamond is commonly referred to as the 'Problem Space' and is also referred to as the 'Research' diamond. Here, exploration, problem synthesis and definition happen. The first diamond ensures that the design process is addressing the 'correct' problem, where creative and innovative ideas are the result of systematic understanding and not something that occurs at random.

The second diamond is commonly known as the 'Solution Space'. Often labelled as the 'Implementation Diamond' or just 'Design,' the second diamond is where the design process realises, understands the operational aspects of, and validates and scales the product. To bring concepts to life, a number of activities, including visualisation, iteration, experimentation, prototyping, and execution, are carried out in this phase.

The 4 D's of the Double Diamond

The Double Diamond is divided into four distinct phases to accommodate the different activities undertaken in the problem and solution spaces. As one proceeds from left to right, each phase has a theme to guide the work, as shown in the diagram below.

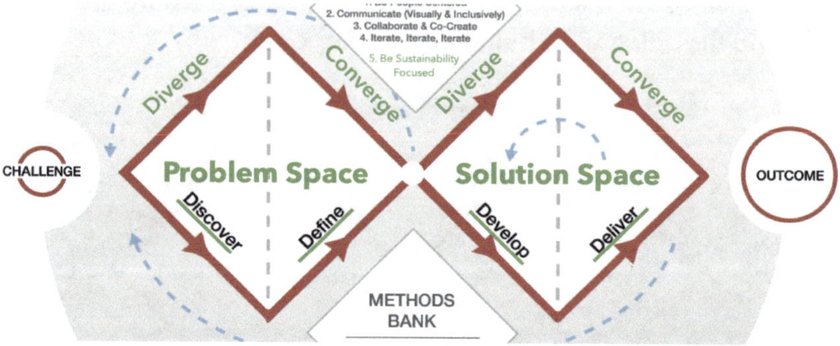

For each phase, the inclusion of sustainability-based activities will help to embed green design into the approach and ensure that these can be consistently considered during problem definition, experimentation, prototyping, iterative testing, and incorporation into the final solution. Let's examine each of these phases and activities in greater depth, and with sustainability in mind.

The Discover phase

In the Discover phase, the aim is full problem immersion to seek to understand the context of the problem and also where exploratory research into the problem might be conducted. Discover focuses on questioning the problem, allowing for dreaming 'big' through the use of divergent thinking. Today, teams are increasingly using experimentation and micro-experimentation as approaches to form testable hypotheses in early-stage discovery. These are good techniques with the same purpose: to better understand the problem before jumping straight to a solution.

Key sustainability activities during this phase should consider the following:

- Planning with stakeholders and identifying the context, including the challenges and opportunities resulting from the long-term sustainability of the product, the environment, and the needs of the customer.
- Conducting primary research with consumers and relevant experts to include considerations for sustainability and environmental impact.
- Utilising secondary or desk research to best inform the team about any sustainability-related angles, considerations, or product requirements.

The Define phase

The objective of the Define phase is to transform the research data obtained in the Discover phase into a workable problem definition or statement. This problem definition will be enriched with data and provide a context that will be useful for the solution work in the next diamond. The process of connecting research data with design ideas is referred to as 'Design Synthesis'.

This second phase employs convergent thinking to synthesise and interpret the data in relation to the problem being solved. The final product will be a 'Design Brief' that summarises and defines the problem and clearly identifies the obstacles that must be overcome. Importantly, this phase can also identify more challenges that can be linked back to further phases of discovery.

Key sustainability activities during this phase are:

- Identifying customer wants versus needs through the lens of sustainability.

- Analysing and synthesising research data and developing insights that will inform decisions and prioritise sustainability.
- Creating a final problem statement that incorporates sustainability and environmental impact and also considers how to measure and track these impacts as part of the design brief.

The Develop phase

In the Develop phase, the goal is to generate potential solutions for the defined problem. It is time to seek inspiration from multiple sources and to share the creative output with as many stakeholders as possible.

The third phase develops multiple solutions based on the design briefs. This is another phase of divergent thinking, which allows for big-dreaming and ideation, but this time within the confines of the brief. During this phase, the various solutions will be prototyped and tested, preferably with colleagues and end users.

Key sustainability activities to consider here are:

- Low-fidelity sketching, wire-framing, and prototyping of solutions, including the development and testing of sustainability-based criteria.
- Collaborating with stakeholders and co-creating with them.
- Developing hypotheses and designing experiments that incorporate the Define phase's sustainability requirements.

The Deliver phase

The goal of the Deliver phase is to validate working solutions and prepare the design for release into the real world. Identifying ineffective solutions through early prototyping and testing at the outset of this phase allows for solution iteration and the preparation of the designed product for its launch.

This final implementation phase concludes with the completion of all design work.

The concluding phase of the double diamond should produce the desired outcome and use convergent thinking to take forward one of the proposed solutions, which will eventually become the launched solution.

Importantly, this phase can identify the need for alternative solutions that can lead to subsequent development phases. It can also reroute back to the beginning of the model if it identifies additional obstacles that require repeating the entire process. As a result, the Double Diamond can be cyclical, with the flexibility to redirect to previous phases when necessary.

Key sustainability activities to consider are:

- Testing prototypes or solutions with customers to obtain early validation of the solution's sustainability impact.
- Iterating and preparing launch assets, including those that demonstrate the solution's sustainability credentials.
- Launching pilots, features, and experiments to learn about the consumer and whether or not the solution meets their expectations for sustainability.

Summary

The sustainability of a digital product or service should be a central component of a broader problem statement and needs to be evaluated alongside the needs and wants of the end user. This will become an increasingly important indicator of the product's success as customer demands for sustainable digital products and services increase.

Using the Double Diamond model is an effective approach to finding the best solution that balances people, profit, and the planet. This so-called triple bottom line is also an excellent basis for making and prioritising design decisions.

To summarise:

- The diamonds in the Double Diamond are the result of divergent and convergent thinking.
- The problem and solution diamond work together to improve outcomes.
- Discover, Define, Develop, and Deliver make up the Four Ds of the Double Diamond.
- There are numerous design process models from which to choose, but the Double Diamond is a powerful approach to creating a solid foundation from which to drive sustainability in design.

The Principles of Green Design

A green design framework is built on both design-based thinking and principles that lead to maximising the energy efficiency of new digital products and services, reducing their carbon footprint, and lessening their overall environmental impact. These principles focus on the reduction of electricity consumption as well as the inclusion of other 'green' factors, such as clean energy-powered hosting and more energy-efficient design choices. The principles should also enable designers to make environmentally conscious decisions while balancing and, ideally, not compromising customer demands.

The framework covered in this section consists of eight key design principles that should be considered and implemented during the stages of problem solving and solution development:

Principle 1: Adopt a Sustainability-Based Design Maturity Model
Principle 2: Build a Green Content Strategy
Principle 3: Optimise Performance
Principle 4: Design a Sustainable UX
Principle 5: Introduce Sustainable Software Engineering
Principle 6:Use Green Hosting and the Cloud
Principle 7: Measure Digital Impact
Principle 8: Make Sustainability a Persistent Constraint

In addition to reducing environmental impact, using green design principles provides a number of additional benefits, which can be summarised as follows:

- Improved performance
- Improved usability and accessibility
- Reduced operational costs

- Supporting the creation of purpose-driven teams
- Improved customer satisfaction

Achieving all of these benefits alongside the development of sustainable digital solutions is a compelling proposition. So let's examine the principles in greater depth.

1. Adopt a sustainability-based design maturity model

At the heart of all effective design approaches is a clear maturity model that supports teams as they become more aware, learn, practice, begin to optimise, and take on leadership responsibilities. This kind of maturity model is even more important when an organisation is attempting to establish a culture of sustainability, as it enables measurement of adoption and identifies the areas of development needed as they progress along their maturity journey.

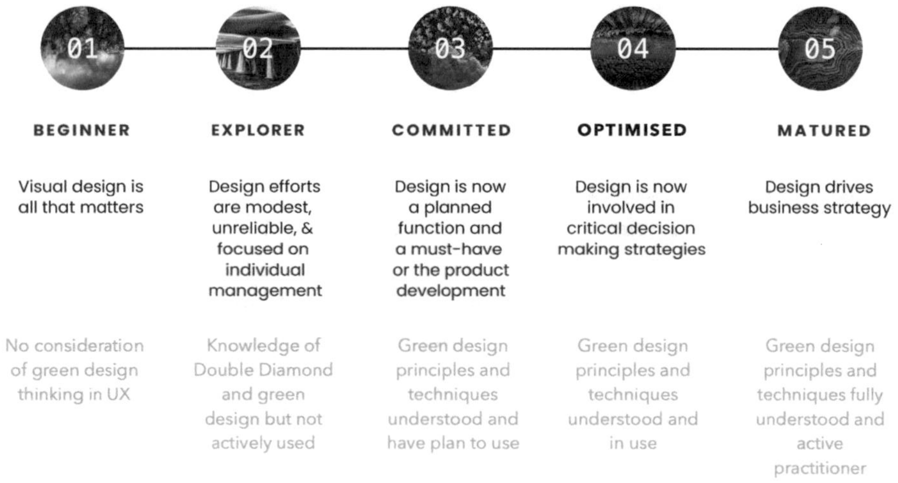

There are many recognised design maturity models, for example, 'The Design Ladder' by the Danish Design Centre, the 'Invision Design Maturity Model', the 'Keikendo Model', and many others, and these generally all use a stepped-level approach. The diagram on the previous page shows how a maturity model can be created to incorporate green design.

By building sustainability into a standard design maturity framework, it is possible to create a stepped-level approach as follows:

Level 1: Beginner

At the beginner level, the focus is almost exclusively on visual design, and there is frequently a gap between what designers and software programmers are attempting to create. At this stage, there is no consideration for developing a sustainable digital solution, nor is its significance recognised.

Level 2: Explorer

At this stage, team members have design responsibilities, but design processes are not fully developed. The design efforts are minor, unreliable, and not prioritised as a necessary approach or enabler. Here, teams function independently and inconsistently. There are pockets of green design, but there is no clear application to projects in flight.

Level 3: Committed

Design is now an integral part of the product development process and a function in its own right. Teams have implemented a design process and a common design system. Committed-level organisations have defined what it means to be a designer, an engineer, and a product manager, and design leaders are now included in the organisation's most important decision-making sessions. Green design is becoming a topic of discussion, and there are emerging plans to incorporate it into problem-solving and solution-finding initiatives.

Level 4: Optimised

At this stage, design decisions are based on data to ensure transparency, and the design process is supported by analytics, experimentation, and initial measurement. All new initiatives heavily rely on customer insights, and any customer-facing activities must receive approval from the design function before moving forward. Executives now incorporate design into their critical decision-making metrics, and design executives are fully engaged and have authority in the process. Green design is an integral component of decision-making, and sustainability outcomes are weighed against return on investment (ROI) and customer needs.

Level 5: Matured

At this level, design governs every aspect of the business, including strategy, vision, and execution. All internal strategies and product development are founded on design-based thinking. In collaboration with the design function, investment decisions for new ideas, features, and products are developed and implemented. Sustainability and green design are central to the organisation's culture, and the design function is regarded as a leader in its field or industry.

I have seen very few organisations achieve higher than Committed level when it comes to incorporating sustainability as part of their design thinking. Now more than ever, there is a need for collaboration between design and software engineering to be able to achieve the green outcomes needed to deliver environmentally-friendly digital solutions. The challenge for organisations is whether they can bring these teams together with a common purpose, and designers have a critical role to play in helping to make that happen.

2. Build a green content strategy

The content being delivered to end customers lies at the heart of every digital product or service. To make the content as sustainably minded as possible, you should create a green content strategy that includes the following:

- An efficient content management process with outcomes that are actionable and transparent to users.
- All content creation workflows should be efficient, collaborative, and deliver results quickly.
- If user action is intended, the content should be readily accessible, clearly organised, and, most importantly, help motivate the user to take action.
- It should be aware of and utilise display patterns relevant to the end user's devices and platforms.
- The content and the efficiency of delivering it to end users should be measurable.

Content Measurement is key

The saying 'you can't change what you don't measure' applies when creating a green content strategy. By selecting the appropriate analytics tools and utilising them to make content choices, it is possible to eliminate waste, streamline processes, and create new content to meet specific demands. However, care should be taken when configuring these tools, as measuring everything isn't always productive, and it's easy to waste a great deal of time measuring things that are irrelevant.

A strong content strategy is based on clearly defining the metrics associated with the success of the product and applying the most suitable methods for achieving the desired results. These strategies help customers locate content more quickly and provide methods for making content more engaging, efficient, and faster to download. This establishes a basis for measuring performance and the use of advanced optimisation techniques that can deliver increased efficiency and reduced energy consumption.

"Perfection is achieved, not when there is nothing more to add, but when there is nothing left to take away."

Antoine de Saint-Exupéry[88]

This quote epitomises the essence of green design. It is a simplified and more environmentally conscious approach to digital design, where, if only the key information is presented, the design is already nearing a perfect state. It goes without saying that reducing content means lowering a product's carbon footprint, as less energy is needed for storage, management, and processing. We should aim to start with this core premise and use it as our baseline before negotiating any add-ons.

3. Optimise performance

The requirement for close collaboration between design and software engineering teams is one of the central tenets of design-based thinking. This relationship becomes critical when trying to optimise the performance of any new product or solution, as it will require a careful balancing act between functionality, speed, reliability, and sustainability.

At present, the balance is not being achieved. The majority of digital design continues to load products with high-definition content. If you open almost any corporate website, you'll see they take advantage of video streaming and high-quality imagery without any apparent loss of speed. These sites are speed-optimised but not necessarily sustainability-optimised. In fact, the

---

[88] From 'Airman's Odyssey', 1942

direction of travel is very much one-way, per the analysis completed by the HTTP Archive:

> **"In the 10-year period between June 2012 to June 2022, the median (web) page weight increased by 221 percent for desktop, and 594 percent, for mobile page loads. "**
>
> Web Almanac by HTTP Archive [89]

Behind these increasing page weights lies a global technical infrastructure needed to deliver this content to users. This infrastructure requires energy to operate, and with this scale of growth, it is unsurprising that the technology sector is projected to consume more than 20 percent of global energy by 2030. It may be difficult to understand how an average web page size increase can result in such a high level of energy consumption, but if you consider the scale of increased internet usage and the global expansion of digital, it all adds up very quickly.

Many small things, when combined, can have a very large impact, and this is why the way in which we design and then optimise all future digital solutions is so crucial.

Managing client expectations

One of the greatest obstacles a designer will face is educating clients who demand high-definition graphics, advanced social features, video, and other add-ons, all of which increase the product's size and energy consumption. Designers will need to purposefully engage clients on the importance of digital product sustainability and guide them towards making 'greener' choices. Using a green design framework as a method for introducing this during the discovery and problem definition stages is a great way to manage compromises and find the right balance together.

---

[89] almanac.httparchive.org 'Annual State of the Web Report', 2022

4. Design a sustainable UX

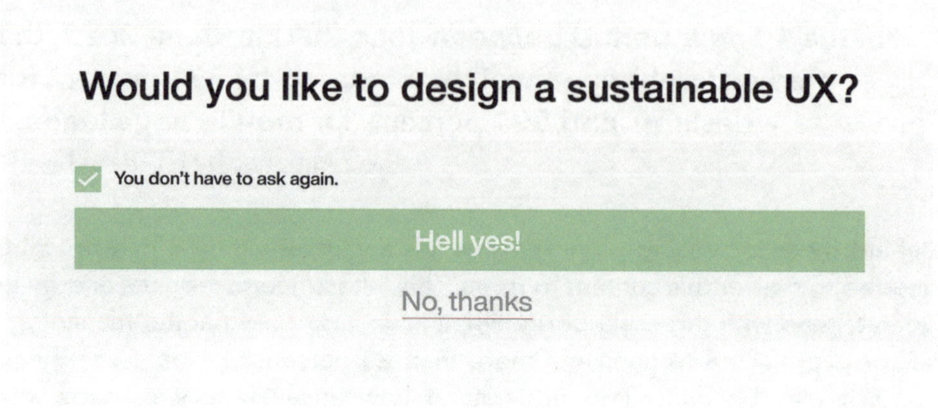

Who in their right mind would answer no?

As a guiding principle, consider sustainability as an essential component of the UX design process and ensure that it forms part of the decision-making and solution criteria for your product. It is no longer enough for UX practitioners to only concentrate on features and design elements that directly impact the immediate end-user experience. The green design framework can provide a method for addressing the complexities of the bigger picture.

When adopting this principle, consider the following four factors when making UX decisions and keep in mind the simple rule: less waste equals more sustainable.

1. Consider the presentation layer

This is the look and feel of the experience; it is what is commonly referred to as its 'visual design language'. Ask yourself if the decisions being made are as efficient as possible. Are they appropriate and proportionate in terms of performance and energy consumption?

2. Consider the infrastructure layer

This relates to the technology that the product uses. Ask whether the choices aided or hindered the UX's sustainability. Are there alternatives or compromises that could be made to significantly affect performance efficiency?

3. Consider the task layer

This describes the application's flow and user interaction. Streamlining user task flow to improve UX efficiency is crucial for developing more sustainable digital products and services. When examining the task layer, consider how to accomplish the following:

- Ability to constantly evaluate the design process itself and improve its efficiency.
- Ensure that the navigation and architecture of the product or service are as user-friendly as possible.
- Assist design teams in designing with actual, rather than mocked, content to understand the true impact of their decisions.
- Ensure that users can complete their tasks in as few steps as possible.
- Consider whether users are enjoying the product regardless of the device or platform they use.
- Determine whether the proposed solution addresses a real-world issue. Is it truly required?

4. The importance of user testing

Incorporating user testing into each stage of the design process is key when designing a sustainable UX. Putting sustainability-focused design decisions to the test with real customers generates transparency and, more importantly, data that can be used to improve design and product choices throughout the process. Using 'A/B testing' methods, in which two UX options are presented to users and their interactions are monitored, can help to provide better and quicker design decisions.

5. Introduce Sustainable Software Engineering

As we've already covered in Part Two, sustainable software engineering is essential to delivering greener digital products and services. During the design phase, the software development teams must be deeply integrated into the problem-solving and solution-creation processes so that early-stage prototyping and subsequent build phases have sustainable engineering approaches embedded from the start.

By building highly efficient applications with lower energy consumption, we can reduce the long-term global impact of the IT infrastructure. When combined with green design, we are able to create digital products that are better for the planet.

Small and large businesses alike are starting to demand software that is simple to deploy, consumes less energy, and occupies less space in the cloud, all without compromising functionality. They are increasingly supporting companies that invest in greener software development and can demonstrate they are transitioning to a digital world that is more efficient and sustainable. Sustainable software engineering can help reduce energy consumption by minimising:

- Energy use during engineering and maximising sustainable code development.
- Resource use on servers and supporting infrastructure.
- Resource use on the client computer or device.

Consider using open source code

Consider the extent to which you can use open source software when evaluating solution options. Why build something that has already been built when you can freely share, use, modify, distribute, and enhance open source software and tools that promote reuse and collaboration? There is also the potential to accelerate solution development and leverage global communities of programmers who are enhancing open source software[90]. Just remember to validate whether the open source code has been designed with efficiency in mind.

6. Use green hosting and the cloud

Any digital product or service that is being designed, prototyped, iterated, built, or used will require energy. Every stage of the new product's life cycle, from ideation to eventual decommissioning, will consume energy, with a significant portion of that energy consumption coming from where it is hosted.

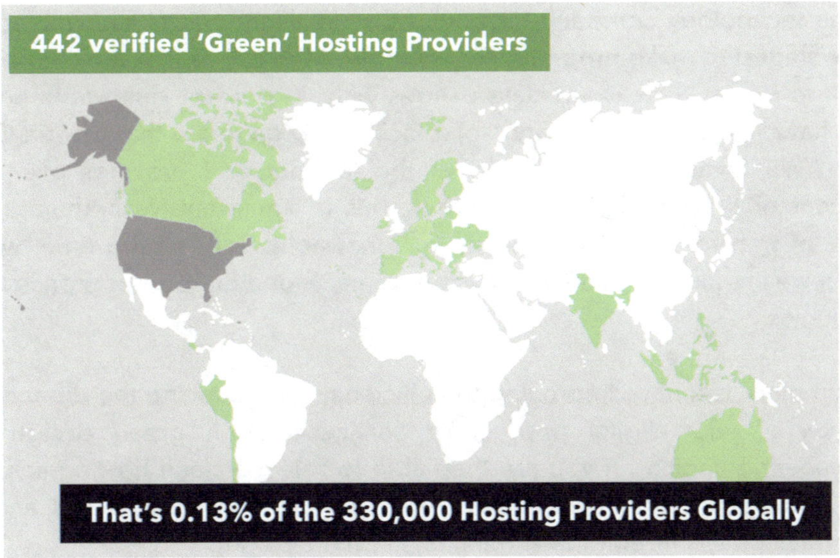

[90] Articles such as 'How to Fight Climate Change with Open Source' at fosslife.org provide more information and links to further resources

The carbon impact and emissions directly related to the digital product or service you are designing and developing can be significantly reduced if you use hyperscale data centres in suitable geographic locations that are powered by a high percentage of renewable energy. However, a Greenpeace audit[91] of data centre providers revealed that the average percentage of renewable energy used to power servers was just over 14 percent. Per the diagram on the previous page, the Green Web Foundation[92] has only verified 442 web hosting companies in 35 countries as 'green' i.e., directly powered by 100 percent renewable energy. That is only 442 out of 330,000 hosting companies worldwide, or 0.13 percent.

The fact is that most compute available is neither clean nor green; there are simply not enough ultra-clean hosting options available at present. So reduction through efficiency and optimised, climate-conscious design choices remains the most viable, measurable, and effective method of enhancing sustainability in the short to medium term.

Some technology companies are investing in direct onsite renewables and have started to make progress. For example, AWS claims that they remain on track to power their cloud data centres with 40 percent renewable energy and have invested in wind and solar facilities to generate that power. Given that AWS hosts up to one-third of the Internet and provides over sixty percent of the world's cloud services, this is significant. Nonetheless, this level of transformational transition to renewables will require time, which, when we are already in a climate crisis, remains one of our most constrained resources.

In summary, prioritise improving the efficiency and reducing the demand for energy of your digital product or service through green design and engineering choices. If you are then able to select a clean hosting solution, then the digital product will already have a reduced energy demand, and the combination will further lower the impact on the planet.

---

[91] Greenpeace, 'Clean Cloud 2021' report

[92] Green Web Foundation and Enterprise Apps Today

7. Measure digital impact

> "If you cannot measure it, you cannot improve it"
>
> *Lord Kelvin*

It is often said that if something cannot be measured, it cannot be changed or improved. And this is never more true than when trying to determine the carbon emissions associated with a digital product or service. Measurement allows us to create a baseline and then understand the impact of actions that can be taken to improve energy efficiency, optimise energy utilisation, and reduce energy consumption.

During the design phase, select a tool that can help you measure the carbon impact of your product design choices and, ideally, simulate the impact before coding. Designers must collaborate with their IT and software engineering counterparts to ensure that the tools selected can deliver consistent, reliable measurements that can also establish a baseline and benchmark a product against similar products. Measurement is essential for setting goals, monitoring performance, and determining if a final product meets its sustainability-based design goals.

I'll cover measurement and tooling in more detail in Part 4.

8. Make sustainability a persistent constraint

Non-Functional Requirements (NFRs) are used to define system attributes like security, reliability, performance, maintainability, scalability, and usability. NFRs (also known as system qualities) are as important as functional epics, capabilities, features, and stories in the agile delivery model. They serve as persistent constraints or restrictions on the system design across the various work backlogs of the product in development. Sustainability, or 'Low Carbon', should now become a key NFR with interdependencies across all other requirements per the diagram below.

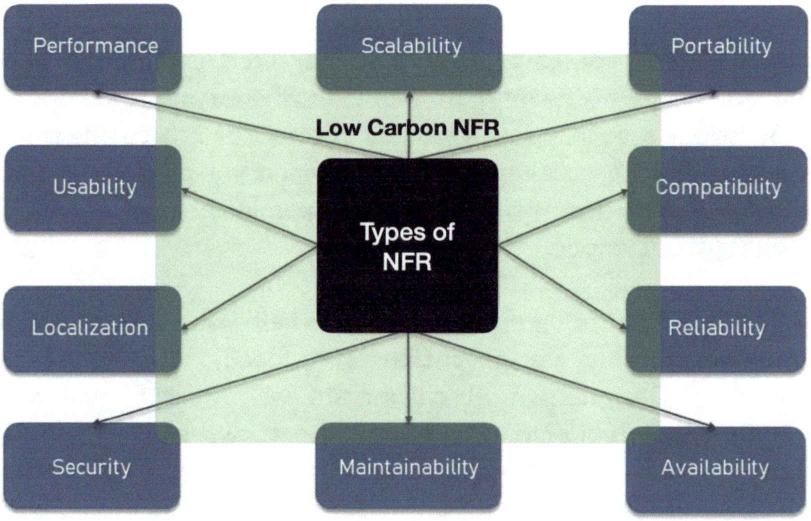

To deliver software products with a focus on sustainability, a 'Low Carbon' specification that describes the system's operational capabilities and constraints and defines how well it will perform alongside standard NFRs, such as speed, security, reliability, and data integrity, is needed. In the diagram, I've elevated the 'Low Carbon' element to reflect how decisions and choices in each of the other NFRs can be made with sustainability in mind.

By incorporating a Low Carbon NFR into the design process, the usability and efficiency of the entire system can be assessed alongside sustainability goals. Failure to meet any of the core non-functional requirements can result in systems that do not satisfy internal business, user, or market demands or that do not meet regulatory or standards agency requirements. In certain situations, non-compliance can result in significant legal issues, such as privacy, security, and safety, and increasingly, with new mandatory financial reporting standards, sustainability is beginning to fall into this category.

The NFR establishes a set of persistent qualities and constraints that will be revisited as part of the sign-off and acceptance criteria for the new digital product or service being created. It becomes a crucial tool for ensuring that the solution meets the design criteria established during the problem-solving and solution-finding phases. The NFR may also impose restrictions on design freedom, so this must be carefully considered during the Discovery phase so as not to inhibit creative thought. However, when used correctly and at the appropriate time in the design process, a Low Carbon NFR can be a very effective means of obtaining your desired sustainability outcomes.

Low carbon NFR in practice

Let's bring this to life in an example. A Low Carbon NFR can help guide decisions, such as reducing image size while maintaining an acceptable user interface experience, while simultaneously balancing the need to meet an agreed target carbon intensity score per software unit as a product requirement. Here, a user-friendly user interface is a functional requirement, whereas a reduced image size is a constraint for achieving the desired Low Carbon NFR. In the acceptability criteria for any backlog item that builds UI functionality, the reduced image size constraint would then be referenced.

In Part 4, I will provide you with access to a Low Carbon NFR specification that you can use on your next project or major product change.

The Techniques of Green Design

So far, I have discussed green design-thinking and highlighted a framework to help guide your design decisions. I've also covered the fundamental principles of green design that will support progress from problem to solution. This final section covers the techniques that will aid prototyping, design iteration, and the building of your new digital product or service.

> "If you think good design is expensive, you should look at the cost of bad design"
>
> *Dr. Ralf Speth*

This statement is a truism, and relevant from both a product development and a sustainability perspective. Going forward, design that ignores the impact on climate and nature should be considered 'bad design' because the environmental costs and consequences can be so devastating.

To help turn green design principles into a practical reality, the following ten techniques should be used to support your design decisions and enhance the sustainability of all new digital products or services:

Technique 1: Simplify the UX and Task Flow
Technique 2: Increase Reuse
Technique 3: Improve Accessibility
Technique 4: Adopt Tooling to Drive Optimisation
Technique 5: Manage Your Use of Colour

Technique 6: Manage Your Use of Images
Technique 7: Manage Your Use of Video
Technique 8: Manage Your Use of Fonts
Technique 9: Understand the Platform and Device Context
Technique 10: Focus on Building Collaboration and Teamwork

1. Simplify the UX and Task Flow

These techniques are especially useful during prototyping, as they can help to avoid unnecessary processing steps and make UX design less complicated. Striking a balance between simplifying the user experience, the user task flow, and sustainability goals is a great outcome for any new digital product.

Simplify the UX

Simplifying the user experience can be achieved through various techniques, including:

- Remove pages and assets that aren't frequently used, as there is no value in retaining pages that perform poorly or are unused.
- Clearly define user requirements by being as specific as possible about what users want and need. Try to find a point of convergence between the two.
- Define content types and understand the content that will help you achieve those goals (for example, using blog posts rather than interactive decisioning), as they can result in a significantly different UX design. Plan to evaluate and use the most effective options.
- Visual design decisions contribute to a sustainable UX; consider using A/B testing to enable direct UX comparisons; test different interface designs to determine if users can complete tasks quickly and efficiently;

test whether navigation is intuitive, and verify that users have a relevant experience regardless of device or browser type.
- Avoid UX patterns that encourage users to increase their viral activity, as this increases energy consumption.
- Enhance the user experience by avoiding proprietary formats and plug-ins and deploying techniques and tools such as progressive enhancement, mobile-first, HTML5, and Cascading Style Sheets (CSS 3).

Simplify task flow

The following techniques to simplify the user task flow can also help maximise resource utilisation and reduce waste:

- Redirect pages to eliminate unused landing pages from the product or service (but be sure to redirect the original URLs to the most closely relevant pages that still exist, as this will prevent search engines from registering 404 errors).
- Actively guide users to complete their work by optimising (1) user flows to ease navigation; (2) search to help users find content with accurate and straightforward results; and (3) clear messaging that helps the user take action.
- Analytics and benchmarking. Simplification is made easier when data is available. Select analytic tools that can help set metrics and track that objectives are being met, and invest in benchmarking to demonstrate improvement and compare performance with original goals. These data points will enable a balance to be struck between task flow, UX design simplification, and sustainability-based product objectives.

2. Increase reuse

Try to maximise reuse across all elements of design, including components, assets, and UI, as this can help promote the rapid development of new products, new variants, and enhancements to existing products without incurring the time and cost of designing and manufacturing from 'scratch' each time. The following techniques should be considered when optimising for reuse:

- Reuse content and assets whenever appropriate and practicable.
- Reuse previously designed software components and design elements that are reliable and well-tested over building new components in order to expedite prototyping and delivery.
- Minimise non-recurring development tasks such as specification, development, testing, and validation with design reuse.
- Try to identify reusable assets that are already known to be energy-efficient.

3. Improve accessibility

Accessibility-based design decisions have been shown to result in increased productivity, optimised performance, and content simplification. All of which can contribute towards a more sustainable and energy-efficient product or service.

> **"When working on digital accessibility, one finds frequently that some practices result in reducing the size of pages, or the amount of data transferred to the user. Furthermore, an accessible Web site is generally simpler and therefore faster to view, not only for users with disabilities. Every small earning humbly contributes to the global effort."**
>
> Open Web Group[93]

[93] 'Accessibility is Good for the Planet', see openweb.eu.org, June 2014

Consider the following when designing for accessibility and sustainability, as when successfully combined, they can result in win-win outcomes:

- Use auditing tools for accessibility to evaluate, understand, and optimise inclusion.
- Use accessible embedded help and online documentation.
- Clarify user data policies with an emphasis on fairness.
- Follow the POUR (Perceivable, Operable, Understandable, and Robust) guidelines.
- Design the product or service to comply with accessibility standards such as W3C and WCAG2[94].
- Ensure that the product can be used by individuals with disabilities and ageing mobile devices.
- Balance design decisions to try and achieve the highest performance with the fewest resources across all devices and platforms.

4. Adopt tooling to drive optimisation

Utilising tools and techniques to evaluate, understand, and optimise the weight of each product feature, component, or page can generate substantial benefits and help support you in making sustainable design decisions.

Consider the following approaches and tools to deliver optimisation enhancements:

- Conversion optimisation is a technique used to evaluate and improve the ability of a page or screen to motivate users to take a desired action.
- Tree testing is used to evaluate navigation structures in order to assist users in locating content more quickly and efficiently.

---

[94] Reference: www.w3.org World Wide Web Consortium: 'Web Content Accessibility Guidelines'

- A/B testing involves providing users with two distinct content or design options and evaluating which performs better.
- Utilising a Plugin Performance Profiler (P3) to determine which plug-ins are taking the longest to launch, how much data they consume, and other performance metrics can help target optimisation opportunities.
- Use software security products and firewall monitors to evaluate the efficiency of the product's security features.
- Use products that can integrate CSS and JavaScript and speed up load times to improve load performance.
- Optimise caching to speed up page load times by bypassing elements that are already cached in the end user's browser or caching files. This can prevent a browser from making repeated server trips for files that are already cached.
- Explore database optimisation opportunities using tools that can reduce the size of a database by optimising its files for faster content delivery.
- Create mobile pages using products that automatically generate stripped-down versions of a page that load more efficiently and faster than the original.
- Remove unnecessary HTML by creating leaner markup and eliminating additional classes to keep code lean.
- Compress and aggregate files using tools to help compress files and execute JavaScript to reduce HTTP requests if the user's browser caches the script.
- Utilise shared libraries and publish static content on a content delivery network (CDN) to reduce the time it takes for users to retrieve content.

These are good examples of optimisation tools and approaches that, when combined with green design and sustainable software engineering techniques, can create a powerful way of decarbonising digital products and services.

5. Manage your use of colour

Less energy is required to power a device in dark mode. Using darker colours in digital design can reduce the energy required to power white or brightly coloured pixels on a display. This is especially true for OLED (organic light-emitting diode) screens, which are now the standard for smartphones and tablets. In dark mode, when displaying a completely black background, screen backlighting is not activated, and none of the diodes will illuminate, thus conserving power.

In addition to dark mode, consider the following colour management techniques for reducing energy consumption:

- Use colour contrast techniques and colour contrast combinations to limit colour use while maximising the effect of basic colours.
- Try to communicate information without relying solely on colour.
- Provide a choice between a light and dark mode so that the user needs to make a conscious decision, knowing the energy implications of their preferences.

6. Manage your use of images

Images can add a significant amount of extra data to the weight of a digital product, and as good practice, you should start by considering whether an image is even required, and if it is, whether the same result can be achieved with a less energy-intensive option such as using resolution-independent CSS effects or web fonts.

When an image is absolutely required, consider using the following techniques to improve the sustainability of your design:

- Image compression: prior to uploading to a content management system (CMS), choose a format that maximises compression rates while balancing image quality; for example, vector image formats such as SVG are resolution-independent and can scale up or down without adding additional weight, whereas PNG or GIF files are appropriate for diagrams or graphs, and JPEG can be a more efficient option for photos.
- Lazy load images: 'Lazy loading' images prevent them from loading until the user scrolls and they hit the browser's viewport.
- Use tools such as ImageCache to further optimise images after they have been published to the CMS.
- Select the smallest image possible to achieve the same visual impact.
- Reduce or remove images that are not visible to the user, such as those stored in carousels.
- Include accessible captions for images.
- Use alternative (alt) text for all non-decorative images.
- Ultimately, try to minimise the number of images and ensure that there is real user value for each image selected.

7. Manage your use of video

"There were over three billion internet users watching streaming or downloaded video at least once per month in 2022. This number is projected to increase annually, reaching nearly 3.5 billion by 2023."

Statista[95]

The use of digital is growing. We are spending more time online and consuming energy-intensive content. Globally, online video is now the most popular digital service, accounting for over 60 percent of all data

[95] Source: Statista, 'Global number of digital video viewers 2019-2023', May 2023

transmitted. It is estimated that video streaming emits as much as 300 million metric tonnes of CO_2 annually, which is equivalent to the total carbon emissions equivalent to a country the size of Spain[96].

Consider the following strategies for reducing the quantity, quality, and energy consumption of video in your design:

- Try to minimise the video length while still conveying the same message.
- Remove autoplay. Ask the user to select play, as this can reduce the number of times the video needs to be transmitted and loaded.
- Include preview images for video playback.
- Include accessible captions for videos.
- In the same way as managing image use, try to minimise the number of videos and ensure that there is real user value for each video that is selected.

Consider using A/B during prototyping phases to determine the user value of video assets, and this can help determine if the impact is significant enough to retain a video or seek alternative designs to meet user requirements.

8. Manage your use of fonts

The screenshot on the next page is of a website by an Amsterdam-based design studio called Formafantasma[97]. It has lots of white space, few images, and only two typefaces: Arial and Times New Roman in standard blue and black. They chose these default typefaces because they have a low data

---

[96] Source: theshiftproject.org, 'Climate Crisis: the unsustainable use of online video', 2019

[97] See formafantasma.com, a collaboration with Studio Blanco

footprint and are probably the most energy-efficient typefaces available for web design.

All content that needs loading on a web page, including fonts, ends up being a request for data. These requests, known as HTTP requests, demand energy and storage space. For a website to be as sustainable as possible, these requests must be as small as possible. Arial and Times New Roman are the default fonts on both Macs and PCs, so no additional requests to download a new font are necessary. It's a simple design choice, requiring a visual tradeoff, but if selected, one that is not only more efficient but can also improve the user experience because pages load more quickly.

When selecting which font to use, consider the following:

- System fonts are installed by default on the user's computer or device and use the least amount of resources, making them the most environmentally friendly option.

- Web fonts provide nearly limitless options and can be displayed on devices where the font is not installed. As they are hosted on remote-to-device servers, the number of HTTP requests increases.
- Using CSS (cascading style sheets), you can embed typefaces into your web page. Although embedding a font results in a balance between an increase in page weight and the number of HTTP requests.
- Use optimisation techniques to remove unused page elements.
- Use no more than two fonts.
- Use SVG instead of web fonts for icons.
- Consider typefaces designed for screen rather than print.
- Only load fonts that are required.
- Reduce the number of font weights and use appropriate font sizes to reduce the font's footprint.

Consider your options carefully when selecting fonts. If the project requires a design with high fidelity, choose a web font. If optimal efficiency is required, consider using a system font. Consider the device context of the end user as well. For example, a web font may work well for desktop users, but a system font will be more effective on a mobile device. Try to ensure that your design decisions don't impede product performance by weighing all font options alongside other techniques already covered.

9. Understand the platform and device context

The platform and device context represent the settings in which a digital product or service will be used. The variables involved can be wide-ranging, including the features, user expectations, differences in operating systems and devices, usage patterns caused by various environments, and differences in the users themselves. Designing for these differences while creating rich and rewarding digital experiences that are sustainably minded can be a major design challenge.

A decade ago, the context of usage rarely changed. Whether the user was in the office or at home using a personal computer, the environment was comparable, and generally no adaptation was required. In today's digital world, smartphones, tablets, laptops, smart TVs, and even the user interface in your car provide access to digital services in very different contexts. Different consumers will use these services at different times and in different places, requiring different interaction models. The best design ensures that the user does not need to consider the device they are using, changes in the environment, or changes in context and can rely on useful functionality and usability all the time, but we have to balance this with the carbon cost of providing these services in these ways.

The following environment, platform, and device considerations can create useful data points to help manage this balance:

- Think about the potential range of platforms and connection speeds covering cloud, edge, and on-premise computing.
- Consider the potential range of devices and connection speeds, including mobile and desktop.
- Assess the different environments and access to technology, such as older devices and less bandwidth.
- Consider the range of different user backgrounds and abilities.
- Consider the variety of potential input methods, such as keyboard, touch, and voice.
- Find out whether digital products and services hosted by providers use renewable energy to power their servers.
- Assess whether the office environment in which your digital products and services are created supports and encourages more sustainable choices.
- Try to understand whether the people designing solutions clearly understand the environmental mission of the product, service, and company creating them.

10. Focus on building collaboration and teamwork

"Alone we can do so little; together we can do so much"
Helen Keller[98]

Product development relies on teamwork and collaboration in the design process because it supports inclusion and brings together a wide range of perspectives and expertise. By leveraging the experiences of a diverse group of people, it is possible to generate more creative and innovative ideas that may not have occurred to a single individual or a homogenous group. Collaboration can also increase the product's overall quality by enabling more comprehensive problem-solving. Involving diverse stakeholders in the design process will result in greater acceptance and support for the final product.

There are a range of methods and techniques that can be used alongside green design to foster collaboration and cooperation. Common ones include:

Multi-criteria decision analysis

A decision-aiding technique to analyse alternatives to complex problems using weighted triple bottom line criteria of people, planet, and profit. The end result of this analysis will generally be the 'best fit' solution.

Thematic dialogue tables

Small-group discussions are held on complex issues and are designed to make decision-making more manageable for participants.

---

98 Not sure of exact quotation source, but possibly 'My Story of my Life', by Helen Keller, 1902

Hackathons

Collaborative computer programming events involve participants engaging in intensive joint ideation to generate solutions and often code a first prototype.

In addition, there are visual design tools that can accelerate collaboration, such as Invision, Figma, and Sketch for UI design and prototyping, Miro and Balsamiq for wire-framing, and Trello for overall project administration and coordination.

As best practice, consider the following:

- Collaborate with software engineering to optimise the choices for compute, storage, and network utilisation.
- Collaborate with the broader business team to ensure social value is balanced with the social and environmental impact of the solution.
- Collaborate with software engineering to estimate the environmental impact of the user experience.
- Collaborate with the business and ESG teams to set measurable sustainability targets.
- Use collaboration tools to help manage and streamline ideation, communication, prototyping, and testing.

Conclusion to Part Three

The EU Science Hub highlights the importance of design in the pursuit of a low-carbon, sustainable digital future. It states that more than 80 percent of all product-related environmental impacts are determined during a product's design phase. Design-based thinking is necessary for creating greener digital solutions. By employing design-based thinking at the outset of every new project, designers, architects, and software engineers can carefully consider the decisions they make, the prototypes they create, and the solutions they deliver in the context of the product's user, customer, service, and environmental context.

A green design framework can help identify the correct problems to be solved prior to coding. The Double Diamond model, which was developed by the UK's Design Council, is a visual model that assists in the creation of value by first identifying problems and then developing solutions. By incorporating the challenges associated with achieving sustainability into the model, we can start to design to reduce climate impact. And by taking into account the user, customer, and service context of a digital product, designers can create solutions that balance amazing features and customer experiences while minimising negative impacts on the planet.

Using green design principles can result in enhanced performance, usability, accessibility, decreased operational costs, high morale, mission-driven teams, and enhanced customer satisfaction. These principles are supported by techniques such as simplifying UX and task flow, increasing reuse, improving accessibility, adopting tooling for optimisation, utilising colour, images, video, and fonts, understanding the platform and device context, and fostering collaboration and teamwork.

Green design can maximise the efficiency of digital applications, reducing their carbon footprint and improving their positive impact on the environment and nature.

Practical - Green Design Checklist

Use the Green Design Checklist on the next page as you consider the sustainability of the digital product or service you are creating. Use the checklist as a reminder of the green design principles and techniques as you go through the various stages of the design process.

You can also access and download the checklist using the QR code above, or alternatively, visit: www.gocode.green/sustainabilityassessmenttools

GoCodeGreen
Decarbonise Digital®

Green Design Checklist

Design-based Thinking foundations		Tick all that apply
Use Design-based Thinking	Understand and apply the concepts of the Double Diamond model to your design thinking at the start of your project	☐
Use the Double Diamond Checklist	Create a Double Diamond Checklist for your project (Discovery-Definition-Development-Delivery)	☐
Divergent and Convergent Thinking in practice	Use divergent and convergent design thinking to create your solution	☐
IT Collaboration	Collaborate closely with IT during the design thinking process	☐
Prototyping	Build a prototype for the project	☐

Sustainability in Design - Principles		Tick all that apply
Sustainability Oriented Design Culture	Establish a design culture and embed design-based thinking to drive sustainability focused problem solving	☐
Content Optimisation	Adopt Content Optimisation principles	☐
Performance Optimisation	Adopt Performance Optimisation principles	☐
Sustainable UX Design	Implement sustainability based considerations for UX design	☐
Sustainable Software Engineering practices	Implement Sustainable Software Engineering principles, techniques and practices in your IT team	☐
Green and Clean Hosting Choices	Select Green Hosting & Cloud	☐
Measurement	Introduce sustainability based measurement into the design process	☐
Sustainability as a Non-functional Requirement	Introduce a Low Carbon Non Functional Requirement into the design, solution and delivery of the product	☐

Sustainable Design - Techniques		Tick all that apply
Simplify UX and Task Flow	Simplify the user experience to increase task based efficiencies	☐
Increase Reuse	Increase the level of reuse of content and assets	☐
Improve Accessibility	Improve the level of accessibility	☐
Adopt Tooling to Drive Optimisation	Introduce tools to evaluate, understand and optimise the weight (data and storage needs) of your solution	☐
Use of Colour	Improve the use of accessible colour combinations and light / dark modes	☐
Use of Images	Efficient image sizing, minimising number of images and use of vector graphics or CSS styles	☐
Accessibility for Images and Videos	Utilise accessible captions for images and videos	☐
Video	Minimise video length, quality and number in the solution	☐
Fonts	Efficient use of font choices and options; only load fonts when needed, and minimise the font footprint and sizing	☐
Platform Context	Ability to adapt to different platforms, network speeds and range of devices where the solution will be deployed / offered	☐
Device Context	Ability to adapt to different ages of device and range of potential input methods for the solution	☐
Collaboration and Teamwork	Improve collaboration and teamwork with software engineering teams to optimise use of compute and set sustainability targets for the solution	☐

Part Four: Leadership that Drives Action

Sustainable IT Leadership

It is now clear to me that achieving the scale and pace of changes needed requires a different type of leadership than we are used to. The traditional leadership traits of being able to do the right things, making the appropriate decisions, interventions, and prioritisation calls, and acting at the appropriate moment to effect positive change all remain true. But taking that first major step, often as a pioneer in your industry, to demonstrate your beliefs by sponsoring an initiative that begins the process of calculating the carbon footprint of technology and digital products and services is needed if you want to be a sustainable IT leader. There can be no more observing the problem. You'll need to focus on getting the data to empower teams and exert pressure on your ICT supply chain; create an environment that supports action; and have faith that people will adopt your vision and develop a plan to mobilise and effect the change we need.

Why Sustainable IT is important

For many businesses, technology has become critical for product development and customer service, forming the very foundation of business success. And as you now know, IT has a sizeable carbon footprint, and technology leaders must do their part to make digital as energy-efficient and sustainable as possible, as any IT-dependent organisation will find it impossible to achieve its Net-Zero commitments without addressing the carbon impact of its technology. As an IT leader, you'll need to build a story that explains why taking action is important for your business. You'll be expected to know both the financial implications and the potential savings that may result.

This represents both a challenge and a significant opportunity for all technology leaders. I sincerely hope that you will choose to accept the challenge; it may prove a defining period in your development as a leader.

The Silent Technology Leader Syndrome

So what kind of sustainable IT leader do you want to be? I now refer to a phenomenon I call the 'Silent Technology Leader Syndrome'. It is based on my interactions with a large number of senior technology and digital leaders over the past two years. Imagine being the technology executive seated at the board table who feels unable to contribute meaningfully to the company's Net-Zero goals; your silence is growing deafening.

I developed a leadership model to help explain the behaviours I've observed. In this model, shown in the diagram below, 'value' refers to the importance placed on sustainability relative to other priorities, whereas 'data' refers to the quality of information available to inform sustainability-based IT decision-making.

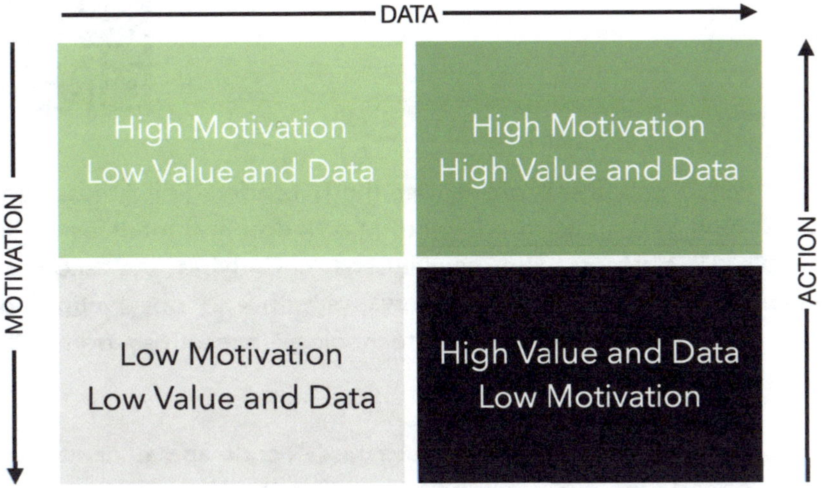

Leadership Positioning Model

The majority of technology leaders currently sit in the upper left quadrant, which I refer to as the 'Silent Technology Leader' box. They are personally highly motivated to contribute to tackling climate change but cannot raise the priority sufficiently in their organisations and don't feel empowered by data to make effective decisions and choices. The challenge is to move all technology leaders into the upper right quadrant, the 'action quadrant'. The remaining parts of this book are written to help you make that transition by developing a sustainable IT strategy to harness your organisation, people, and business behind a common goal: to decarbonise digital.

Practical - IT Leaders Self Assessment

Let's start with a practical exercise. Use the IT Leaders Self Assessment and receive a score indicating your level of knowledge and readiness to lead a sustainable IT strategy. The calculations underlying the scoring and evaluation system are based on interviews with and observations of the behaviours of a large number of technology executives from different organisations and industries.

You can access the self assessment using the QR code above, or alternatively, visit: www.gocode.green/sustainabilityassessmenttools

Please answer honestly

The score you receive is intended for self-reflection and to help highlight learning areas that will be useful as you read through the remainder of the chapter.

Key Terminology for Leaders

I want to start by building on the terminology explained in Part One. It is essential that leaders have a firm grasp of key climate change terminology, such as the Greenhouse Gas (GHG) Protocol scopes 1, 2, and 3; the ICT Sector Guidance; and an appreciation of specific calculation approaches and methods, as well as a more detailed understanding of the definitions of Net-Zero and Science-Based Targets. All of these terms are now part of common business vocabulary and are important for understanding how the climate challenge across technology is being framed.

A recap of scope emission categories

As previously discussed (remember the weekend barbecue session?), the Greenhouse Gas Protocol developed the scope 1, 2, and 3 methodologies to represent the various categories of greenhouse gas emissions. Businesses and organisations are under increasing pressure to first measure and then explain how they will reduce all three scopes.

Scope 1 emissions

These are 'direct' emissions, which a business generates by operating the assets it possesses or controls. These can result from the operation of machinery used to manufacture goods, the use of vehicles, the heating of buildings, and operating information technology. From a sustainable IT perspective, selecting infrastructure components and hardware devices that operate with greater energy efficiency could help reduce scope 1 emissions.

Scope 2 emissions

These 'indirect' emissions result from the energy a company purchases. Installing solar panels or procuring renewable energy rather than using electricity generated using fossil fuels would reduce a company's scope 2 emissions.

Scope 3 emissions

These are also indirect emissions in the sense that the company does not produce them, but they differ from scope 2 in that they also include emissions from clients who use the company's products or suppliers who make goods the company uses. Scope 3 has the concept of 'upstream' emissions, those that come from the production of a business's products or services, and 'downstream' referring to emissions related to their use or disposal.

There are 15 categories of scope 3 emissions as shown in the list below and contextualised alongside scope 1 and 2 emissions in the diagram on the next page[99]:

- Purchased goods and services
- Capital goods
- Fuel- and energy-related activities
- Transportation and distribution (upstream and downstream)
- Waste generated in operations
- Business travel
- Employee commuting
- Leased assets (upstream and downstream)
- Processing of sold products
- Use of sold products
- End-of-life treatment of sold products

---

99 Source: brightest.io

- Franchises
- Investments

For the vast majority of organisations, scope 3 is the most difficult to address as it can be challenging to obtain, making scope 3 emission measurement more complicated than for scopes 1 and 2.

Measuring scope 3 emissions necessitate monitoring activities across the entire business model, or value chain, from suppliers to end users. It will require time, standards, common practices, collaboration, and a commitment to transparency and honesty in order to achieve the depth and breadth of measurement required. How we incrementally measure, improve, and accept the fact that data will change and not always be 100 percent accurate will be a technology leader's challenge.

Scope 3 is also frequently the most significant source of emissions for a business, accounting for more than 70 percent of an organisation's carbon footprint and sometimes reaching up to 90 percent. Companies can normally more readily measure their scope 1 and 2 emissions and can control them by taking steps like switching to renewable energy or electric vehicles. However, Scope 3 emissions are under the control of suppliers and consumers, so they are impacted by decisions made outside their direct control.

When I started GoCodeGreen, I remember a very early conversation with a friend who also happened to be the head of ESG at a major European bank. After explaining the importance of measuring digital products and services to enable decarbonisation, she asked me, 'I like the idea, but how big is the IT problem? I need to know if it is material to my overall emissions, because if it isn't, then I'll focus elsewhere'. It was a good question, and one I can now answer.

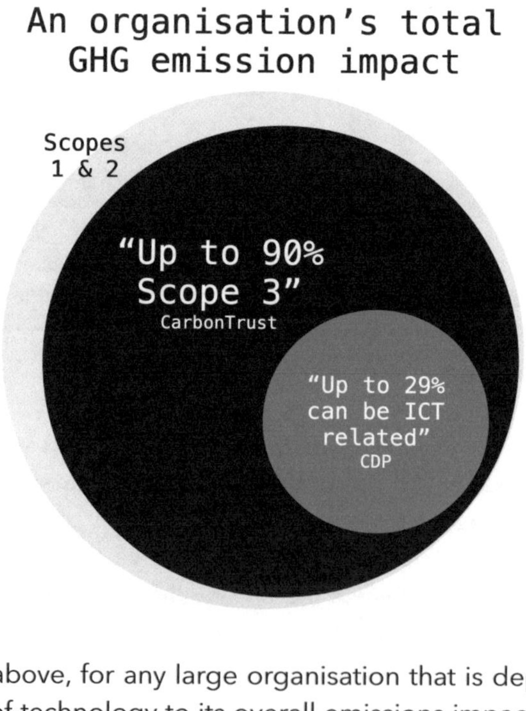

Per the diagram above, for any large organisation that is dependent on ICT, the contribution of technology to its overall emissions impact can be as high as 29 percent[100]. The Shift Project and Carbon Disclosure Project undertook

---

[100] Sources: Carbon Trust (carbontrust.com); Carbon Disclosure Project (CDP) - cdp.net ; and shiftproject.org

analysis at the IT product level across a large sample of businesses across sectors to map these scope 3 related emissions. More recently, I have seen these figures play through from undertaking assessments in some of the world's largest organisations.

So the level of ICT impact for a business is material. And this is going to become more of a challenge as Net-Zero commitments must include reductions in scope 3 emissions, despite the fact that they are beyond its direct control. Organisations are now becoming aware that they must understand the scale of their ICT-related impact and need help to address this. I'll cover the solutions and approaches to tackling these challenges later in the chapter.

All technology leaders should make a concerted effort now to understand all of their ICT and digital scope impacts and start to develop strategies to mitigate them.

Other key climate related terminology

In addition to the GHG Scope definitions, I've found the following terms useful to understand when speaking with IT leaders and ESG professionals: Some are a recap from earlier sections, but useful reminders nonetheless:

Location vs. market based calculations

A 'location-based' method uses the average emissions intensity of power supply grids where energy consumption takes place (using grid-average emission factor data for the most part). Whereas a 'market-based' method reflects emissions from electricity that businesses have chosen to purchase. This allows a company to use and report their purchase of green energy tariffs via Renewable Energy Certificates (REC) and Guarantees of Origin (REGO) without necessarily using the renewable energy in their operation.

Materiality

Materiality assessments are formal exercises designed to determine the importance of environmental, social, and governance (ESG) issues to stakeholders. In a digital world, technology and IT are increasingly recognised as one of the material elements that must be measured and acted upon by many organisations.

Gross vs. Net Carbon

The term 'Gross' encompasses emissions from all sources at every stage of a product's lifecycle, whereas 'Net' emissions are subject to interpretation regarding what is excluded, such as embodied emissions, the use of renewable energy sources, and emissions minus offsets. Gross carbon calculations, taking into account existing data limitations, provide the best starting point for action.

Cradle-to-gate /-grave

'Cradle-to-gate' evaluates and measures the carbon emissions of a product until it leaves the factory (gate) and is ready to be delivered to the consumer. 'Cradle-to-grave' incorporates all life cycle stages in a measurement, providing a complete environmental footprint from start (cradle) to finish (gate).

Embodied vs. operational carbon

'Embodied' carbon is the carbon footprint of a material or product. It considers how many greenhouse gases (GHGs) are released throughout the supply chain and is often measured from cradle to factory gate or cradle to site of use. Embodied carbon can also be measured along the boundaries of cradle-to-grave, which is the most complete boundary condition. This boundary includes the extraction of materials from the ground, transport, refining, processing, assembly, use, and finally at their end of life.

'Operational' carbon is the term used to specifically describe the emissions of carbon dioxide and other global warming gases during the in-use operation of a product.

Emission factors

An emission factor (EF) quantifies the rate at which a particular activity emits greenhouse gases (GHGs) into the atmosphere. Additionally known as conversion factors, emission intensity, and carbon intensity.

ICT Sector Guidance

The GHG Protocol ICT Sector Guidance provides guidance and accounting methodologies for calculating greenhouse gas emissions associated with ICT products and services. The ICT Sector Guidance is based on and conforms to both the GHG Protocol Scope 3 standard and the Product Standard. This may be slightly sad, but it has become the most important document in my life for the past 2 years.

Based on the diagram below[101], the GHG Product Life Cycle Accounting and Reporting Standard can be used to understand a product's full life cycle emissions, thereby focusing efforts on the greatest GHG reduction opportunities and the ability to target these reduction actions at each life cycle stage.

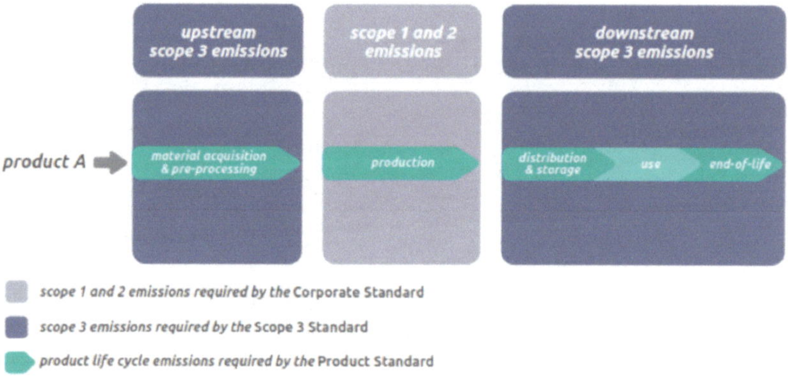

In terms of coverage, the ICT Sector Guidance is based on a series of core chapters, as follows:

- Telecommunications Network Services
- Desktop Managed Services

---

101 Source: GHG Protocol ' Product Life Cycle Accounting and Reporting Standard'

- Cloud and Data Centre Services
- Hardware
- Software

The ICT Sector Guidance recognises that assessing the GHG emissions of technology products presents a number of challenges due their complex and extensive characteristics, the lengthy and complex supply chains for IT hardware, and the extensive use of shared resources within ICT systems, which requires more complicated allocation techniques. The ICT Sector Guidance attempts to resolve these issues by providing practical methodologies that can be used to calculate the GHG emissions associated with ICT goods and services in a consistent manner.

Many manufacturers and IT service providers are starting to assess their own emissions impacts and, increasingly, those of their products; however, the absence of standards means that trust levels surrounding their claims are frequently perceived as biassed (to promote sales) or inaccurate (as calculated emissions can vary by orders of magnitude between similar products and services).

At GoCodeGreen, we have the capability to consistently measure across the five main categories in the GHG Protocol ICT Sector Guidance. Using the tools I've developed, it is possible, for the first time, to consistently measure and understand your organisation's full use of ICT across scopes 1, 2, and 3 and at the digital product level to drive reduction, action, and efficiency. This measurement provides the context to then use workload and code-level tooling, power monitors, and profilers to target lower-level, detailed analysis when required.

Over the next few pages, I'll describe how I approached tackling this challenge, and over the remainder of the book, I'll show you why IT leaders now have access to the framework and measurement capabilities that will allow them to decarbonise digital. This marks the end of the Silent Technology Leader.

Chasing down carbon impact

Three years ago, when GoCodeGreen did not exist and I was an innovation lead in financial services, I sat in a room of UK bank executives at the beginning of a new product launch. I suggested that we design and develop it to be the bank's greenest ever application. I recall the table nodding and leaving the room wondering, 'How am I actually going to do that?'. I quickly realised that I would need a sophisticated measurement tool that could help me measure the carbon impact of digital. However, despite an extensive search, I was unable to locate anything that would meet my needs. I decided to experiment, formed a hypothesis, and started to create the solution I was looking for.

My hypothesis: software is at the top of the ICT hierarchy; if you can make an application more energy efficient, then you can reduce the demand for the entire ICT value chain that supports it.

From early on, I was clear on what I did <u>not</u> want to create, which is sometimes a good place to start. I didn't want to design a source code analyser or cloud workload measurement tool. Instead, I wanted to approach the problem from the perspective of driving action and impact. I wanted to create an elevated, organisation-level context for digital decarbonisation with software as the focal point for taking action. In my view, software is at the top of the ICT hierarchy; if you can make applications more energy efficient, then you can reduce the demand for the entire ICT value chain that sits beneath it.

I also wanted to create a tool that, in itself, would not create even more carbon emissions. Running real-time analysers across software and hardware devices that would consume even more energy just didn't feel like the right answer.

I use the two diagrams on the next page to illustrate to clients how GoCodeGreen can help. First, I ask them to imagine being able to understand the carbon impact of the ICT that supports their businesses, their digital products, and services across the various dimensions of desktop managed services, telecoms and network managed services, data centres, and the cloud (i.e., the chapters of the GHG Protocol ICT Sector Guidance). Then to imagine the hardware devices sitting in each of these three categories of ICT being screened and the total carbon emissions and energy usage being estimated within acceptable levels of tolerance and accuracy. This would then provide the all-important context from which to derive action through software.

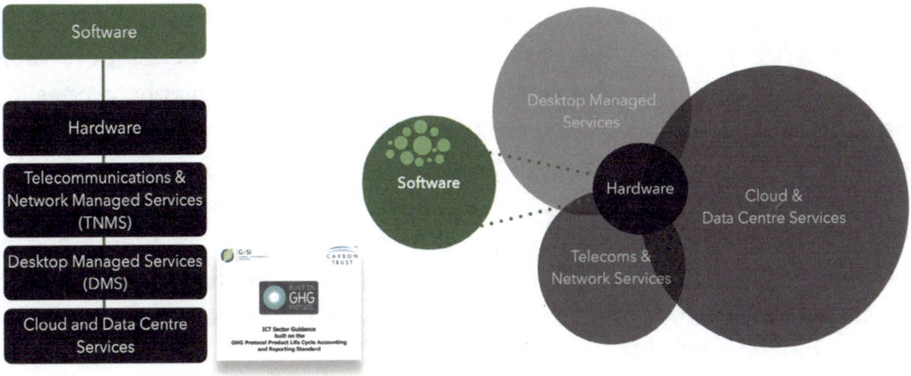

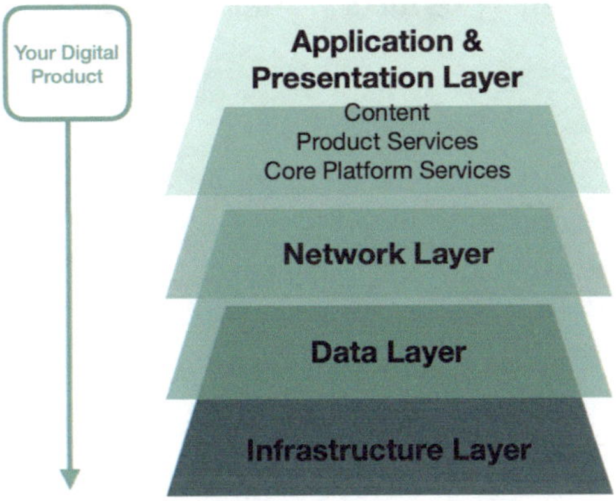

Then I ask them to imagine that the software and applications (digital products) that drive the demand for those hardware components and the associated energy usage can also be calculated, with everything being tracked through the architecture layers, baselined, and rated, and the actions to help reduce the energy demands of the software and associated hardware can be connected via an underlying calculation methodology that is consistent across the entire architecture stack. It is this holistic view across an organisation's ICT with digital product-level detail to drive targeted action through the layers of the technical architecture that I created at GoCodeGreen.

Finally, I ask clients to imagine being able to aggregate results across multiple digital products so that we can identify the scaled decarbonisation levers across an entire ICT estate.

This is the approach I like to take: decarbonising business and customer outcomes by mapping the software and architecture that support value chains or customer journeys within an organisation. I ask clients to consider the value of developing carbon neutral digital products and services and, by doing so, how this might increase their sales and contribute to their growth. I explain that by understanding the emissions related to their own products, they will be better placed to respond to new reporting standards and to requests from their own customers wanting to understand their scope 3 emissions. I always conclude by explaining how all of these actions can result in reducing the cost of IT.

I wish I could have used capabilities like this three years ago.

If you want to do this for the ICT and digital products in your organisation, we should talk.

A Leaders View on Net-Zero

As we know, to prevent the worst effects of climate change and preserve a habitable planet, the increase in global temperature must be limited to 1.5°C above pre-industrial levels. The Earth is currently approximately 1.1°C warmer than it was in the late 1800s, and emissions continue to increase. In accordance with the Paris Agreement, emissions must decrease by 45 percent by 2030 and cease altogether by 2050 in order to keep global warming below 1.5°C.

Net-Zero means cutting greenhouse gas emissions to as close to zero as possible, with any residual emissions being reabsorbed by oceans and forests. By decarbonising digital and tackling the emissions related to ICT, IT leaders have an important role to play in helping their organisations achieve Net-Zero.

In order to achieve Net-Zero, leaders must:

- Prioritise rapid and deep emission reductions by establishing near-term targets, i.e., reducing direct and indirect value-chain emissions, to limit global temperature rise to 1.5°C. The goal of halving emissions by 2030 is the top priority for all businesses.
- Establish long-term objectives, i.e., attaining Net-Zero, requires companies to establish long-term, science-based objectives to reduce all possible emissions by 2050. The majority of businesses must reduce their emissions by more than 90 percent. To attain this, the Science Based Target Initiative (SBTi) has developed the Corporate Net-Zero standard (we've covered this at a high level already, and we will explore it further in the next section).
- Neutralise residual emissions, i.e., after a company has achieved its long-term goal and reduced emissions by >90 percent, it must use permanent carbon removal and storage to offset the final 10 percent of non-eliminable residual emissions. An organisation is only considered to have

achieved Net-Zero status once it has met its long-term, science-based goal and neutralised any remaining emissions.

Action not offsetting

To achieve Net-Zero, the call to action requires us to reduce emissions by more than 90 percent while allowing for carbon capture and storage to offset less than 10 percent of residual emissions. Net-Zero commitments cannot be met through offsetting, either directly or indirectly through the supply chain. The diagram below depicts the various strategies for attaining Net-Zero in order of preference.

Approaches to achieving Net-Zero

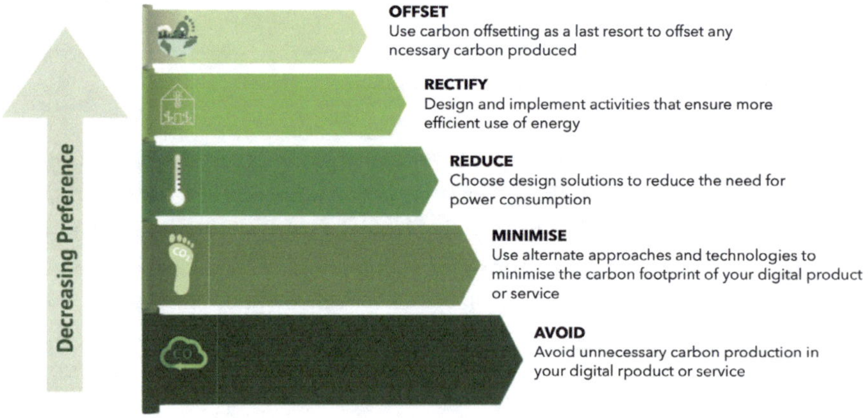

Avoidance, minimisation, and reduction are the preferred and only truly sustainable ways to achieve Net-Zero, and the same applies to digital decarbonisation. It will be these three approaches that underpin the creation of a sustainable IT strategy discussed in the remainder of this chapter.

Please stop offsetting

The Science Based Targets Initiative's Corporate Net-Zero Standard[102] states plainly:

> **"Carbon credits do not count as reductions towards meeting your science-based targets. Companies should only account for reductions that occur within their operations and value chain."**

Transitioning to a Net-Zero world is one of humankind's greatest challenges, and avoiding, minimising, and reducing current energy consumption levels is essential to achieving it. Offsetting and the use of carbon credits should not replace your direct action and cannot be seen as the answers to the challenges we face.

Customer and Investor Demand

Consumer demand

Companies need to care about the environment if they want to keep and attract customers. An increasing number of consumers base their purchasing decisions on climate action. According to a study of more than 10,000 people in 17 countries[103], consumers all over the world increasingly base their purchasing decisions on environmental sustainability.

Consumers increasingly view themselves as agents of climate action, and they believe and anticipate that for-profit businesses should also be taking

---

102 SBTi Corporate Net Zero Standard, V1.1, April 2023

103 Simon-Kucher.com, 'Global Sustainability Study 2021', Oct. 2021

climate action. This increases the pressure on business and technology leaders to prioritise sustainability. According to the same study, 34 percent of consumers are now willing to pay more for environmentally friendly alternatives (remember Bill Gates' concept of a 'green premium' in Part One), so damaging profitability can be avoided if companies can reinvent the products and services they offer to these consumers.

Investor demand

Investors can be the stakeholders who exert the most pressure on businesses to care about the environment and take action. Since COVID-19, the interest of investors in climate-related issues has increased, resulting in an increased focus on environmental issues.

According to a survey by EY in 2021[104], 74 percent of institutional investors are more inclined to divest from companies with poor environmental sustainability and ESG records. The demand for improved climate disclosure is increasing, and businesses must respond by becoming more rigorous about materiality, aligning better with finance, and understanding climate risk disclosures.

For technology businesses, especially FinTech and start-ups, who are seeking investment, the ability to demonstrate your ESG credentials is already important. Increasingly, investors and customers are also demanding to see the environmental credentials of your products to make better-informed investment, funding, and procurement decisions. Being able to showcase that you understand and then have a strategy to reduce carbon emissions and environmental impact may well be key to the success of your future business.

---

[104] ey.com, '2021 EY Global Institutional Investor Survey' - 'Is your ESG data unlocking long-term value?', Nov. 2021

Science Based Targets Initiative (SBTi)

Science based targets show businesses how much and how quickly they must reduce their GHG emissions to prevent the most serious impacts of climate change and can help put them on a direct path to decarbonisation. By guiding businesses through science-based target setting, SBTi enables them to combat climate change while maximising their competitiveness in the transition to a Net-Zero economy.

When establishing short- and long-term goals for achieving Net-Zero, a number of factors, including boundary setting, level of ambition, timeline, and methods, must be evaluated. Within SBTi, reduction-based action is central to achieving both short- and long-term goals.

The SBTi provides:

- In accordance with climate science, the definition and promotion of best practices in emissions reductions and Net-Zero targets.
- Companies with methods and guidance for setting science-based targets in accordance with the most recent climate science.
- A team of experts to provide companies with independent assessments and validation of targets.

Science based targets provide a clear framework for managing the ambition of the 1.5°C target. The table[105] on the next page shows the methods for achieving the targets within the required timeframes.

---

[105] Source: SBTi Corporate Net-Zero Standard (reference point 9 pertains to the inability of the maritime transport sector to set a target year for near-term goals prior to 2030).

		Boundary *What percentage emissions inventory coverage is required?*	Ambition *What is the ambition level of limiting temperature rise?*	Timeframe *What is the timeframe to meet targets?*	Methods *What are the eligible methods to set targets?*
Near-term SBTs	Scope 1 and 2	95%	1.5°C	5-10 years[9]	Cross-sector absolute reduction Sector-specific absolute reduction Sector-specific intensity convergence Renewable electricity
	Scope 3	If >40% of total emissions, 67% coverage	Well-below 2°C		Cross-sector absolute reduction Sector-specific absolute reduction Sector-specific intensity convergence Supplier/customer engagement Scope 3 economic intensity reduction Scope 3 physical intensity reduction
Long-term SBTS	Scope 1 and 2	95%	1.5°C	2050 latest (2040 for the power and maritime transport sectors)	Cross-sector absolute reduction Sector-specific absolute reduction Sector-specific intensity convergence Renewable electricity
	Scope 3	90%			Cross-sector absolute reduction Sector-specific absolute reduction Sector-specific intensity convergence Scope 3 economic intensity reduction Scope 3 physical intensity reduction

For IT leaders, the following distinction is important:

SBTis are not set for specific actions related to achieving sustainable IT; instead, think of delivering sustainable IT as being one of the key levers to achieve the Net-Zero commitments of an organisation that will be measured against SBTi targets.

The Impact of Digital

Over the past decade, the number of internet users worldwide has doubled, and global internet traffic has increased by a factor of twelve. Right now, more than half of the world's population (5.18 billion) uses the Internet[106], with more than one million new users logging on daily. We live in a digital world.

According to the Shift Project's report[107] on the environmental impact of digital technology, the ICT sector is now responsible for between 4 percent and 6 percent of greenhouse gas emissions and is projected to consume up to 20 percent of global energy by 2030 if current growth trends continue unchecked.

"The digital advances made are staggering when we look at the short time scales of their technical implementation. But

[106] Statista, 'Internet and Social Media Users in the World 2023', Aug. 2023

[107] theshiftproject.org 'Environmental Impacts of Digital Technology: 5 Year Trends and 5G Governance'. Feb. 2023

they also come at a cost, as the production of the hardware and the electricity needed to fuel this Internet explosion both leave a large environmental footprint."

UN Environment Programme

The UN Environment Programme report, 'The Growing Footprint of Digitalisation' (November 2021), sums up the impact of digital perfectly; here I quote them in full: "Computers, servers, and other electronic devices require large amounts of natural resources. The energy to run them emits high amounts of CO2, and programmed obsolescence and the low percentage of recycling are generating e-waste. The vast majority of data in the cloud is not used. The 'life in the cloud', with our music, photos, movies, emails, documents, and social networks stored on distant servers, accessible instantaneously from a PC, laptop, or mobile phone from almost any corner of the earth, has become the new norm. In addition, most of our economic transactions have become digitalised. It is estimated that 60 percent of global GDP will be digital and that 70 percent of new value created in the economy over the next decade will be based on digitally enabled platforms." Without a doubt, we live in a digital world.

Your company's digital carbon footprint

Carbon impact can be measured in terms of a 'footprint'. This footprint refers to the greenhouse gas emissions produced by digital technology resources, tools, and platforms. The environmental costs of the internet and digital technology are very real, despite the fact that they are frequently out of sight and can therefore be left out of mind.

To understand the digital carbon footprint of your company, consider the following sources:

- Utilisation of devices and infrastructure
- The emissions generated during their manufacture.
- The power required to operate devices and equipment

- The transfer of data
- The amount of power consumed by servers and data centres

The following digital activities are among the most significant contributors to both our global and personal digital carbon footprints:

- Business, financial, and operation processing
- Digitalisation of public sector services (including health and core government services)
- Corporate intranets and websites
- Digital customer services
- The use of cloud computing
- Information and news websites
- Shopping online
- Multi-media
- Online gaming
- Social media
- e-learning and educational services
- Video streaming
- Artificial intelligence and generative AI
- Digital and crypto currencies
- The Metaverse

Assessing a digital carbon footprint can be challenging, especially as so much of the digital world is interconnected in ecosystems, linked data processing, and long, complex supply chains. This makes ICT a hard-to-abate sector, and it will require IT leaders to really think through their approach to decarbonisation. To start this process, complete the practical on the next page and get an idea of your own company's digital carbon footprint.

Practical - Screening Tools

The majority of technology leaders lack adequate awareness of their IT's carbon footprint. When workloads are running in a public cloud environment, a more detailed understanding of the cost and estimated carbon impact for specific workload types may be available, but generally, this only forms part of a much bigger picture. In complex multi-cloud and on-premises environments with a blend of legacy, partially modernised, internal, and third-party IT solutions, it is much more difficult to have a comprehensive and consistent baseline view.

Use the high-level screening tools to gain a first view of the impact of your ICT and the technology choices you have made. There is a corporate version for IT leaders in larger organisations and a SME version for IT managers in smaller enterprises. The corporate version is more suited for those managing large, complex IT estates, frequently with multiple geographical locations and many teams. The SME version is geared towards creating awareness and helping individuals in small to midsized companies take action. These high-level estimation tools will provide you with an idea of the scale of the challenge and opportunity in front of you.

You can also access the Corporate and SME versions of the screening tools using the QR code above, or alternatively, visit: www.gocode.green/sustainabilityassessmenttools

What rating did you achieve?

In the Corporate version, 'High' and 'Very High' ratings for 'Compute and Application Estate' and 'People and Location' are not unexpected. Many IT choices made in the past were focused on efficiency, cost, risk, and other traditional technology decision criteria. They did not consider carbon impact or energy efficiency, so it should come as no surprise that some, if not most, of these choices perform poorly when viewed through a lens of sustainability. The good news is that past decisions that may have resulted in greater carbon impacts can now be measured so that improvements can be made. If we are to make the use of technology more sustainable, we must now focus on remediating the past and improving the new digital products and services of the future.

In the SME version, you will have been sent the 'Low Carbon IT Manifesto' to help identify some of the main causes of technology related carbon emissions in your business as well as actions to reduce them. You can also try using the SME version at the team level in a bigger organisation to see how smaller changes can have an impact when aggregated.

You now know more about your own readiness and the environment in which you operate. Let's start to build on this over the course of the next few sections and explore ways in which you can create and execute an effective sustainable IT strategy.

Defining Sustainable IT

'Sustainable IT' is a broad term that describes an environmentally conscious approach to the design, use, and disposal of computer hardware and software applications and should also include a redesign of accompanying business processes when needed. Often, 'sustainable IT' and 'green IT' are used interchangeably, but there is an important distinction to be made between the two terms.

While green IT refers to an approach where an organisation seeks to reduce the operational environmental impact of its IT estate through actions that increase energy efficiency, such as optimising server usage, transitioning to more efficient physical IT assets, or by seeking to deploy more efficient coding or software, sustainable IT expands the scope to consider the full lifecycle of digital assets and can include raw material use, social justice and human rights, and end-to-end product lifecycle management. A sustainable IT strategy can enable the deep decarbonisation of an organisation's digital capabilities.

Sustainable IT is now an important consideration within the scope of Environmental, Social and Governance (ESG), and its significance is made greater by the increasing reliance of many organisations on technology.

Many technology leaders and teams have started to incorporate sustainability into their agendas, particularly as CEOs and boards have prioritised this critical issue as part of their ESG strategy and publicly stated Net-Zero commitments. But before IT can reduce its carbon footprint and become more sustainable, technology leaders must first assess and understand their current emissions.

What aspects of IT does sustainability encompass?

All aspects of IT should be included in scope, i.e., all the components that make up an IT infrastructure: data centres and cloud equipment, including services, storage systems, networks, and telecommunications systems; including the networks that allow remote users to access corporate systems; and all end-user devices, including desktop computers and peripherals, laptops, smartphones, and tablets, as well as the applications they run to support business operations.

The gap

For many technology leaders, there's a gap between wanting to be sustainable and being sustainable. Matt Bradley, Group Environmental Sustainability Lead at Capgemini, and his team described the situation and the dilemma for IT leaders perfectly in their 2021 report, 'Sustainable IT: Why it's time for a Green Revolution for your organisation's IT':

"Fifty percent of companies said they have an organisation-wide sustainability strategy, but only 18 percent have a complete sustainable IT strategy with clear goals and timelines. 33 percent of organisations mandate IT carbon footprint reporting, but most don't require their vendors to disclose carbon footprint data. Many organisations find measuring the IT carbon footprint to be a complex and difficult task, with 49 percent citing the lack of access to common measurement standards and ratings and the difficulty of understanding the many layers of IT's footprint."

Capgemini

The good news for technology leaders is that the difficulties of measurement and coverage are now being addressed, presenting a significant opportunity to bridge the gap and start tackling this challenge.

Cloud, an Important Part of the Answer

'Move to the cloud' is the often-said solution in many sustainable IT discussions. But let's be clear: the cloud does not miraculously eliminate all the environmental effects of IT, and as recently reported in the media, vendors' claims of Net-Zero are not always as they seem. Numerous industry observers assert that the major cloud service providers rely too heavily on carbon offsets and credits and do not provide enough transparency or granular carbon reporting.

> **"In 2019, Amazon committed to achieving 'Net-Zero carbon' across its businesses by 2040. The online retail behemoth's company-wide emissions have soared by 40 percent since then, topping 70 million metric tons of carbon dioxide last year. It's a glaring example of the gulf between corporate pledges and climate progress."**
>
> MIT Technology Review, 2022

Ultimately, the cloud relies on data centres, and data centres are full of hardware equipment using energy and almost constantly being cooled. All of this requires lots of electricity and water. According to an article in Time Magazine[108], in 2019, Google needed more than 2.3 billion gallons of water to cool its data centres in three US states, with severe environmental and nature-based implications for local natural ecosystems.

In a recent client assessment, GoCodeGreen calculated that an application migrating from an on-premise data centre to a hyperscale cloud service provider resulted in a 12 percent reduction in carbon emissions. This was

---

108 Time, 'The Secret Cost of Google's Data Centres: Billions of Gallons of Water to Cool Servers', April 2020

based on an unusual scenario where the grid carbon intensity between the existing and new data centre locations was almost identical and covered a straightforward workload migration without application modernisation or cloud-native development. 12 percent is a worthwhile like-for-like improvement, but it looks small in comparison to the 90 percent+ efficiency claims being made by some cloud providers. Proper measurement and context-based measurement, specific to your product and organisation, are the only ways to target action and avoid making claims that could otherwise be considered inaccurate, or worse, as greenwashing.

Technology leaders must explore how they manage their data centres and network service consumption and concentrate on utilising the cloud correctly by modernising their applications and adopting sustainable software engineering and green design practices. Only by doing so can cloud efficiencies be genuinely unlocked to assist businesses in enhancing their sustainability. At GoCodeGreen, we've also seen the other side, with clients investing in true cloud-native, modern approaches to engineering their digital solutions and achieving high efficiency and low carbon ratings.

Cloud computing, and energy-efficient cloud computing in particular, can have multiple positive effects resulting from economies of scale, improved workload and compute utilisation, and when powered directly by renewable energy or low-carbon intensity grid locations. In a perfect world, all digital products and services would be able to run in these efficient, clean computing environments, and all consumers would be able to weigh their options alongside other factors such as proximity to customers, regulatory requirements, risk, resilience and stability, and, of course, cost. However, we do not live in a perfect world, and many larger organisations will face a multi-year journey to achieve this goal.

A 'mishmash architecture'

When discussing the complexities of managing large IT organisations with clients, I frequently use the diagram on the next page to illustrate the wide range of software application types and underlying IT architectures that the majority of large organisations have to manage daily. I call this a 'mishmash architecture' consisting of legacy, on-premise application estates that continue to underpin many of their business and operational processes, as well as internal, private cloud environments and usually a much smaller proportion of fully cloud-native, modernised applications on the public cloud. Frequently, the complexity is further exacerbated by decisions to adopt a multi-cloud strategy to manage risk. As a result of these complexities, the opportunity to quickly achieve large carbon emission reductions can be severely diminished, but the scale of reductions possible is vast.

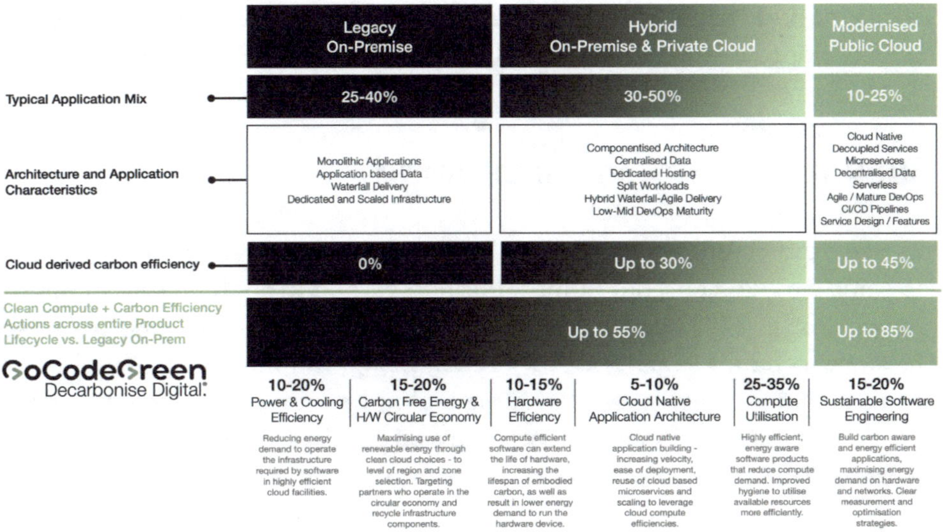

I have seen some businesses decide to migrate from on-premises computing facilities to a public cloud environment without considering the true carbon impact of their decisions. Knowing the carbon grid intensity of energy being used is crucial to understanding whether a cloud migration will actually deliver carbon reduction; for instance, migrating workload from an on-premise data centre in Southern Scotland to a cloud data centre in London could result in the use of energy with three times the carbon intensity (source: National Grid ESO[109]), i.e., you've migrated to the cloud but your emissions may have actually increased!

And as many technologists are aware, a cloud migration exercise can often require more effort and time than initially anticipated, as unravelling the complexity of legacy applications takes time. A simple workload migration without transforming an application to be cloud native so that it can take full advantage of containerisation, microservice based architectures, and elastic and on-demand compute will not deliver the scale of benefit you might expect (or have been sold). Fortunately, most IT leaders are aware of the range of benefits that modernisation can bring, including speed and velocity of delivery, lower cost, improved functionality, and a better user experience. So the added value of doing the job properly to release sustainability benefits alongside everything else should be easy to support and prioritise.

Back to my recipe for change

We need the cloud. And we need the cloud to run on renewable energy in the same way we need our homes, businesses, and everything else to run on renewables. It will happen, but it will take time. While we wait, we need efficiency to be a key part of every action plan, from the country level down to the digital product level. Better managing how much we are consuming and making less stuff that needs energy or requires more energy to create are sensible actions we can take now. Remember my recipe for change?

It is the only way I can envision achieving the required reductions in the short time that we have. Relying on the cloud alone simply won't do. For digital, we need a sustainable IT strategy and a framework for moving the software and infrastructure architecture from the left to the right of the earlier diagram, gaining efficiencies through the course of that journey. And it must be supported by solid, consistent measurement so that the impact of changes can be properly understood.

---

109 See nationalgrideso.com and their 'Carbon Intensity Dashboard'

The Circular Economy and ICT

What is the circular economy?

The circular economy is a production and consumption model that encourages sharing, leasing, reusing, repairing, refurbishing, and recycling of existing materials and products for as long as possible. This extends the life cycle of products and reduces waste to the bare minimum.

The circular economy is an evolution in the global production and consumption of both products and services. The circular model - see the diagram above - reimagines the economy based on the elimination of waste and pollution and aims to keep products and materials in circulation for as long as feasible. In this model, the restoration of the world's wildernesses, the construction of regenerative agricultural systems, the use of renewable materials, and the transition to renewable energy sources must occur concurrently if everything is to become genuinely sustainable.

The shift to a circular economy

The circular economy means abandoning the current, wasteful economic model of the world, which is largely founded on the idea of 'take, make, waste', in which resources are extracted, transformed into products, used, and discarded. But in order to accomplish this shift, products will need to be redesigned to be more durable, reusable, repairable, and recyclable, with the goal of keeping them in circulation for as long as possible. It also requires changing the way in which we consume and utilise products and services and ultimately rethinking consumerism as a society.

The challenge is that our progress has been limited, as the Circularity Gap Reporting Initiative (CGRi)[110] discovered:

> **"The global economy is now only 7.2 percent circular. The global situation is getting worse year on year, driven by rising material extraction and use. Rising material extraction has shrunk global circularity: from 9.1 percent in 2018, to 8.6 percent in 2020, and now 7.2 percent in 2023. This leaves a huge Circularity Gap: the globe almost exclusively relies on new (virgin) materials. This means that more than 90 percent of materials are either wasted, lost or remain unavailable for reuse."**
>
> CGRI, 2023

[110] The Circularity Gap Report 2023

Linear vs. circular economy

The term 'linear economy' is used to describe our current global economy, in which we extract resources, manufacture goods, use them, and then dispose of them. It is based on mass production and consumer demand for short-lived, highly disposable products like technology devices, fashion items, household machines, and cars.

The linear economy has thrived because it provides manufacturers with high profits and consumers, primarily in developed nations, with low prices. The linear economy does not account for the cost of the system's waste, pollution, and carbon emissions.

The reality is that businesses are incentivised to waste, pollute, and emit, as they are rarely required to pay the full environmental or social costs of these actions. If a price were placed on social and environmental damage caused by emissions or waste and simultaneously more circular and sustainable approaches were rewarded, multiple benefits would result, including healthier ecosystems, safer working conditions, and far better environmental products and services.

What are the benefits of a circular economy?

Circularity reduces both the demand for and environmental impact of extracting raw materials. The circular economy can reduce harmful activities such as carbon emissions, air pollution, and exposure to toxins and increase beneficial activities such as habitat restoration, renewable energy, and cleaner air.

Circularity also has the potential to produce substantial economic benefits. The expansion of reuse, repair, remanufacturing, and recycling can generate millions of jobs and stimulate innovation. Capturing and reusing essential materials, such as rare earth metals, makes the economy more resilient to disruptions in the global supply chain and ensures that the world has the necessary materials to construct a renewable energy infrastructure. In every dimension, the concept of circularity works.

Circularity in ICT

The World Economic Forum (WEF) reports[111] that the circular economy for ICT equipment has the potential to reduce the annual production of 50 million metric tonnes of e-waste. The WEF report states:

> **"We need a system that functions properly – in which the circular economy replaces the linear. It is time to reconsider e-waste, re-evaluate the electronics industry and reboot the system for the benefit of industry, consumer, worker, health of humankind and the environment. On average, six kilograms of electronic waste was generated per person globally. This is equivalent to 125,000 jumbo jets".**

With more than 100 billion metric tonnes of earth's resources entering the global economy each year, the circular economy for ICT equipment needs a massive acceleration to limit our impact, reduce our dependence on raw material (or 'virgin') resources, and ensure that these resources are reused.

Promoting the circular economy for ICT requires systemic change, focused primarily on how we consume and manage resources. It is easy to see how the cycle of 'take, make, waste' fits with the throwaway, disposable manner in which we've regarded technology over the past few decades. The following four areas are key to overcome if ICT is to be able to adopt circular principles:

---

[111] World Economic Forum, 'A New Circular Vision for Electronics : Time for a Global Reboot', Jan. 2019

1. Limit our consumption of raw materials

According to a report[112] by the United Nations Environment Programme (UNEP) International Resource Panel, "Global annual extraction of materials grew from 22 billion metric tonnes in 1970 to 70 billion metric tonnes in 2010." According to the CGRi, this level of exponential growth has now reached more than 100 billion metric tonnes annually[113]. Our consumption of natural resources is at double the rate of population growth. We are consuming more than nature can sustain.

More focus must now be placed on repurposing previously extracted materials across ICT. Massive recycling programmes are required, and as a starting point, the world's largest corporations, who are so effective at selling us new stuff, should be held accountable for rapidly pioneering these initiatives to slow down the pace at which we extract raw materials.

2. Be smarter with materials

The issue with many types of ICT equipment is that they were never designed to be repaired, and when businesses are compelled to invest in new technology to remain competitive, demand for new products only continues to increase.

This has the effect of creating excessive raw material consumption, and we enter a vicious cycle: as technological use continues to increase, so does the demand for raw materials to sustain it. We must design more sustainable ICT products, and if we are unable to do so, then we must influence markets by purchasing products that are shown to be more environmentally friendly. Rather than purchasing new models when older ones break or reach the 'end

---

[112] United Nations Environment Programme, 'Global Material Flows and Resource Productivity', 2016

[113] Circle Economy, 'The Circularity Gap Report 2021'

of life' phase of their lifecycle, ICT products should be designed to make repairs possible, thereby creating new markets for spare and replacement IT parts. And we must design and build the software that runs on these devices to be more efficient and able to maintain performance and functionality on ageing or lower-spec hardware.

3. Systemic change

The challenge for most organisations is that their growth and profitability are largely driven by increasing consumption of their products and services, and until recently, with little regard to sustainability, nature, society, or climate. However, on a finite planet, growth cannot be infinite using traditional economic models. These models must be disrupted, and systemic change must be implemented within our global infrastructure. This is why the triple bottom line of balancing people, planet, and profit is so effective and why so many of the world's leading corporations have invested in the B-Corp movement.

At a broader level, there are policies and practices emerging to create a more sustainable economy, such as the political demand for a post-COVID green new deal[114], individual consumer trends, and businesses aiming for

---

[114] See commission.europa.eu for details of the European Green Deal:
"Climate change and environmental degradation are an existential threat to Europe and the world. To overcome these challenges, the European Green Deal will transform the EU into a modern, resource-efficient and competitive economy, ensuring:

- no net emissions of greenhouse gases by 2050
- economic growth decoupled from resource use
- no person and no place left behind

The European Green Deal is also our lifeline out of the COVID-19 pandemic. One third of the €1.8 trillion investments from the NextGenerationEU Recovery Plan, and the EU's seven-year budget will finance the European Green Deal."

'zero waste'. ICT manufacturers and digitally-driven businesses are beginning to respond, but we still have a long way to go before we can declare victory. Governments can also do more to impose duties on products that use only virgin materials or digital products with no or low carbon impact credentials. Incentives like this are required to change the system.

4. The impact of ICT e-waste

The following statistics from the TES Sustainability Report[115] highlight the impact of e-waste and make it clear why action is required:

- In 2016, the world generated 44.7 million metric tonnes of e-waste, and by 2021, that number would rise to 52 million metric tonnes annually.
- By 2040, it is anticipated that 14 percent of our annual carbon emissions will be attributed to the production of electronics.
- According to the UN's Global E-waste Monitor Report, precious metals such as gold and platinum worth at least $10 billion have ended up in landfills.
- Globally, e-waste accumulation is inconsistent. In 2019, the e-waste production per capita in Northern Europe was approximately 23 kilogrammes. In Africa, this equated to approximately 3 kilogrammes.
- Recycled metals can be up to ten times more energy-efficient than virgin metals.
- In 2017, the raw material value of e-waste was $62.5 billion per year.

According to the same report, the renewable energy transition alone will only reduce global greenhouse gas emissions by 55 percent. The remainder will come from how efficiently we produce and consume products and food, as well as how we manage land; this is where the circular economy comes in. We urgently need the world to connect the development of the circular economy to the UN Sustainable Development Goals (SDGs) and link funding to circular outcomes. By focusing equally on environmental and social goals,

---

[115] See tes-amm.com, 'Sustainability Report, 2021'

the circular economy can contribute to the reduction of inequalities, the improvement of well-being, and the enhancement of environmental quality.

The Principles of Circularity

There are three central principles that support the implementation of circularity. When explained, they are like a jigsaw piece, so clearly needed when missing but so clearly part of the solution when found. Over the next few pages, let's discuss how the principles fit together to help create a more nature-oriented economy.

1. Eliminate waste and pollution

The circular economy's first principle is to eliminate waste and pollution. As stated previously, the majority of the global economy currently operates on a 'take-make-waste' model. We extract raw materials from the earth, transform them into goods, and then dispose of them as refuse. The majority of this waste winds up in landfills or incinerators, where it is lost. Because the resources on our planet are finite, this system is defective and cannot function over the long term; it is the very definition of unsustainable.

For many products on the market, there is no onward path after use; they have been created to be disposable and incorporate waste into their design. Our economies are filled with products that were designed without regard for what happens at the product's end-of-life. But most products can and should be circular through maintenance, sharing, reuse, repair, refurbishment, remanufacturing, and recycling as a last resort. With the proper focus during the design phase, we can eliminate the concept of waste. This reinforces the need for green design-thinking to build for and design sustainability into all products as a first principle. And this includes digital products and services.

2. Circulate products and materials

The second principle is to circulate products and materials at their highest value. This involves keeping materials in use, either as a finished product or, if that is no longer feasible, to deconstruct the product into components or raw materials. In this way, nothing is discarded, and the intrinsic value of products and materials is preserved.

This can be accomplished in two ways: the technical cycle and the biological cycle. In the technical cycle, products are reused, repaired, remanufactured, and recycled. Biodegradable materials are returned to the earth through processes such as decomposition and anaerobic digestion as part of the biological cycle.

Technical cycle

For digital and ICT, the technical cycle is really important. Maintaining and reusing products rather than breaking them down into their component parts or raw materials is the most effective method for preserving their value. As a result, the first stage of the technical cycle focuses on keeping products whole to retain their maximum possible value. This may include the design of business models based on sharing, in which consumers gain access to a product rather than owning it, allowing more people to use the product over time. Alternately, reuse might involve reselling combined with cycles of upkeep, maintenance, and renovation.

When a product can no longer be used, its components can be remanufactured. Parts that cannot be remanufactured can be disassembled into their component parts and recycled. Recycling is essential as the final stage that keeps materials in the economy and prevents them from becoming waste. Recycling is often viewed as the last resort due to the loss of embedded value in products and components. Even digital products can be viewed in this manner; for instance, before decommissioning software at the end of its life, consider how to publish and reuse any code assets and data. The hardware that supports a digital product can also be treated in the technical cycle and go through the stages of use, reuse, remanufacture, and recycling.

Biological cycle

Biodegradable materials that cannot be reused, such as certain food by-products, can be circulated back into the economy through the biological cycle. By decomposing or anaerobically digesting organic materials, valuable nutrients such as nitrogen, phosphorus, potassium, and micronutrients can be used to help regenerate the land so that more food or renewable materials such as cotton and wood can be grown.

You are probably thinking, 'the biological cycle doesn't relate to ICT'. My friends at 'Grow Your Own Cloud[116]' have something to say about that. I'll leave you with a taste of what they're doing in the image below. And yes, that is a GoCodeGreen digital report encoded into the DNA of a plant.

116 You need to check out growyourown.cloud and the work that Monika Seyfried and Cyrus Clarke are doing. It's mind blowing.

3. Regenerate by nature

The third principle of the circular economy is nature regeneration. By transitioning from a linear 'take-make-waste' economy to a circular economy, we can better support natural processes and give nature more room to thrive.

From extraction to regeneration

We can transfer our focus from extraction to regeneration, and instead of continually degrading nature, we invest in creating 'natural capital'. We can implement practices that allow nature to regenerate and increase biodiversity, as well as return biological materials to the earth. If we adopt a regenerative model, we will begin to emulate the way in which natural systems have self-regenerated for billions of years.

More space for nature

By keeping products and materials in circulation, less land is needed for sourcing raw materials, such as the mining of silicon to produce the base material for computer processing chips. More and more land can be returned to nature, and rewilding can occur. To achieve this, we must continually decouple economic activity from material extraction by keeping materials in circulation after use.

In a circular economy, land devoted to material sourcing becomes increasingly focused on renewable resources grown in a regenerative manner, as opposed to the extraction of finite materials, which will remain in circulation for longer periods of time. All of this will be supported by a transition to 100 percent renewable energy generated by infrastructure designed for reuse, repair, remanufacturing, and recycling. All of this will require time, which is why an immediate reduction in energy consumption is required. We must help create capacity in the system and support the development of a circular economy and the practices and technologies required to support the transition to renewable energy.

Mother Nature

Let’s talk about trees

"Trees are symbols of hope, life and transformation. They’re also increasingly touted as a straightforward, relatively inexpensive, ready-for-prime-time solution to climate change."

Science News[117]

"There aren’t enough trees in the world to offset society’s carbon emissions – and there never will be."

The Conversation[118]

Two differing viewpoints. Which should we believe?

"I don’t plant trees. It is complete nonsense to think that tree planting could solve climate change. Are we the science people or are we the idiots?"

The last quote was from Bill Gates onstage at *The New York Times*' recent climate summit[119]. Balance and perspective are important here. Trees are good. There is no reason why we shouldn't plant trees to replace the ones we've already cut down or to replace some of the forests lost to agriculture in order to feed our expanding populations, especially if we select the right mix

---

117 Science News, ‘Why planting tons of trees isn’t enough to solve climate change’, Carolyn Gramling, Earth and Climate Writer, July 2021

118 The Conversation, Bonnie Waring, Grantham Institute - Climate Change and Environment, Imperial College, London, April 2023

119 The New York Times, Climate Forward Live Event, ‘Discussing the Threats of a Rapidly Warming Planet’, Sept. 2023

of tree species suitable for the damaged habitat (i.e., the ones we cut down in the first place). However, it can take a tree up to 40 years to absorb one tonne of carbon dioxide equivalent. It's simple logic: not cutting down more trees has a greater effect than planting new ones.

When it comes to removing carbon dioxide from the Earth's atmosphere, we need trees. Through photosynthesis, they extract carbon dioxide from the air to aid in the growth of their leaves, branches, and roots. Additionally, forest soils can store enormous quantities of carbon. Advocates of tree-planting projects say they achieve two vital things: (1) they help offset current emissions, and (2) they draw out CO2 emissions that have lingered in the atmosphere for decades or longer.

The Bonn Challenge[120] is a programme with a global goal to bring 150 million hectares of degraded and deforested landscapes into restoration by 2020 and 350 million hectares by 2030. If we allowed natural forest land that has been lost due to human activity to regrow, it is estimated that these lands would store approximately 42 billion metric tonnes of carbon by 2100. However, if the land were to be filled with single-tree commercial crop plantations, as is typical for tree plantation offset programmes, carbon storage would decrease to approximately 1 billion metric tonnes. Critics of the Bonn Challenge say that these single-tree initiatives constitute a major element of its restoration projects[121].

The intention is good, but the execution and choices need to be carefully managed to ensure the maximum value and impact from the effort and investment. Poorly designed initiatives that do not address a broad range of scientific, political, social, and economic concerns may do more harm than good, wasting money as well as political and public support. There are other concerns too, for example, that there is too much focus on the number of

---

120 bonnchallenge.org, Forest Landscape Restoration

121 Annals of Forest Science, 'Implementing forest landscape restoration under the Bonn Challenge: a systematic approach', Stanturf, J.A., Kleine, M., Mansourian, S. *et al. May 2019*

seedlings planted and too little time focused on how to keep the trees alive in the long term, or that indigenous communities are not engaged when planting on their lands, and that not enough attention is paid to the fact that different types of forests store vastly different quantities of carbon.

It might take forty years before a tree grows to maturity and is able to sequester a meaningful amount of carbon. That is the reason why GoCodeGreen chose to plant kelp in the ocean[122] to compensate for the residual carbon in our business and digital products. Kelp grows three times faster and sequesters up to twenty times more carbon than terrestrial forests. Whether tree planting or growing kelp, neither should be considered an alternative to investing in other energy efficiency or carbon reduction actions but should be used in tandem with them.

To reinforce this point, Carolyn Gramling of Science News says that if we maximised the quantity of vegetation all land on Earth could support, we would only store enough carbon to compensate for approximately ten years of greenhouse gas emissions at current rates. After that, natural carbon capture would not increase any further.

So where do we go from here? As we begin to take action and help create time for new technologies to develop and for broader global climate policies to align, I believe there are two factors we must consider: (1) immediate energy efficiency-related action must take priority, and (2) we must balance our current overreliance on trees with a broader nature conservation strategy and preserve other carbon-storing ecosystems.

In response to reading Bill Gates view on trees, I made the packet shown on the next page and sent it to Bill in response to his Climate Week speech. I'm not sure if he'll respond, but if he does, I'd also like to ask him to open the door for us to work with Microsoft to help them measure and decarbonise the impact of their digital products on their clients.

---

[122] We partnered with SeaTrees to reforest the ocean, sea-trees.org

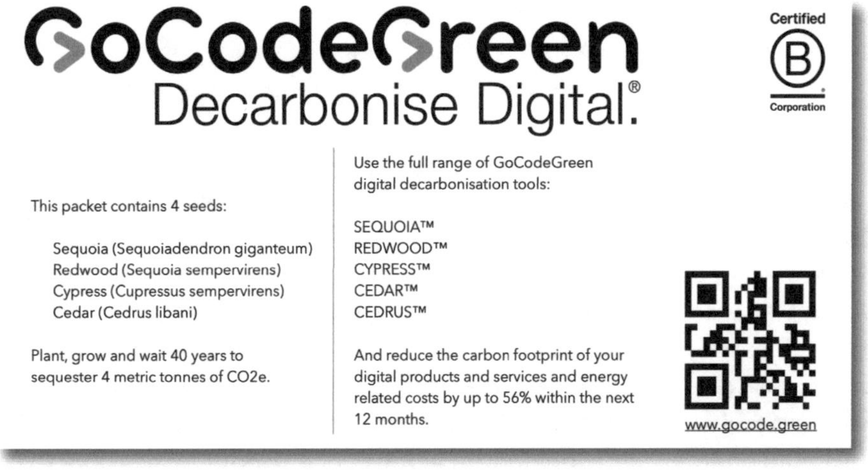

Energy efficiency as well as tree planting

We must stop defining our actions only in terms of offsetting or nature-based compensation. We must consider how to increase energy efficiency and reduce demand by making better choices. Consider the following statement:

> **"We consider energy efficiency to be the 'first fuel' as it still represents the cleanest and, in most cases, the cheapest way to meet our energy needs."**
>
> IEA[123]

Energy efficiency offers some of the quickest and most cost-effective actions to reduce greenhouse gas emissions and will be crucial for closing the gap between climate ambitions and current Net-Zero strategies. IEA Executive Director Fatih Birol continues, "We consider energy efficiency to be the 'first fuel' as it still represents the cleanest and, in most cases, the cheapest way to meet our energy needs. There is no plausible pathway to net zero emissions

---

[123] Fatih Birol, IEA Executive Director, Nov. 2021

without using our energy resources much more efficiently. A step change in energy efficiency will give us a fighting chance of staving off the worst effects of climate change while creating millions of decent jobs and driving down energy bills."

In addition to well-developed energy efficiency policies such as electrical appliance standards, which in some countries have prevented electricity consumption equivalent to their total wind and solar power generation, the IEA also stresses the growing importance of digital technologies for the future of energy efficiency. Rapid adoption of digitally connected devices contributes to expanding the scale and scope of energy efficiency's benefits and can facilitate a cheaper, simpler, and more cost-effective transition to renewable energy. This represents a wonderful opportunity not only to create digital products and services that enhance first-order energy efficiency by substituting traditional energy-intensive activities but also to create second-order benefits by making digital capabilities themselves cleaner.

We've $&%#@! the natural carbon cycle

"The planet naturally releases and absorbs far more carbon dioxide than humans emit by burning fossil fuels. The problem is that human activities have thrown the Earth's carbon cycle out of balance."

MIT Climate Portal[124]

According to the MIT Climate Portal, the planet's natural carbon cycle circulates a vast quantity of carbon dioxide (CO2). The oceans and forests absorb and sequester carbon dioxide for hundreds to thousands of years, and they are referred to as natural carbon sinks. Hydrothermal vents and

---

124 MIT Climate Portal, 'How much carbon dioxide does the Earth actually absorb?', Jan. 2022

subsea volcanoes are two examples of natural CO2 sources that also release carbon. Through this natural cycle, the Earth annually absorbs and emits approximately 100 billion metric tonnes of carbon dioxide.

So far, land plants and the ocean have taken up about 55 percent of the excess carbon people have put into the atmosphere, and as stated at the start of the book, about 45 percent has stayed in the atmosphere. The land and oceans will eventually absorb the majority of the additional carbon dioxide, but up to 20 percent may persist in the atmosphere for many thousands of years. But if humans emit only a tenth as much CO2 as the natural world, why are scientists so concerned that our emissions are causing climate change?

It is because our additional carbon emissions have thrown out of balance what was once a balanced cycle.

"All of this extra carbon needs to go somewhere."

NASA Earth Observatory[125]

Human activity is accelerating CO2 emissions and creating an imbalance. Even though the natural carbon cycle self-balances, it does so over extremely long timescales. And although humans have added our own emissions on top of natural carbon sources, we cannot speed up the work of most of the natural carbon sinks that absorb CO2 from the air. It takes centuries for carbon dioxide in the atmosphere to absorb into the oceans, and approximately another 10,000 years are required for natural processes to remove excess carbon from the oceans and restore their equilibrium.

"It doesn't have anywhere to go and it's not going to go away for a long time."

Professor Daniel Rothman, MIT Department of Earth Atmospheric and Planetary Sciences

---

125 Nasa Earth Observatory, 'Effects of Changing the Carbon Cycle', 2011

A sizeable portion of the carbon dioxide that humans have emitted over the past few centuries will remain in the atmosphere for many years to come due to the extremely slow rate at which natural carbon sinks absorb CO2. It would take hundreds or thousands of years for the planet to rid itself of the excess CO2 that humans emitted during the industrial era, even if they stopped emitting all greenhouse gases tomorrow. This is why scientists are so interested in carbon capture solutions and why we need to stop adding to the problem as much as possible.

Changes in the carbon cycle have an effect on every sink. Extra carbon in the atmosphere heats the planet and promotes plant growth on land. Excess carbon in the ocean causes the water to become more corrosive, endangering marine life. Terrestrial and marine ecosystems play a crucial role in climate regulation; currently, they absorb just over half of all carbon emissions produced by humans. Biodiversity and natural ecosystem services aid in climate change adaptation and mitigation, and they are therefore a crucial part of our fight to combat climate change. We must work with nature rather than against her in order to preserve a liveable planet.

Our impact on nature

Current research[126] estimates that permafrost in the Northern Hemisphere contains 1,672 billion metric tonnes of organic carbon. If only 10 percent of this permafrost were to melt, it could discharge enough carbon dioxide to raise global temperatures by 0.7°C (1.3 degrees Fahrenheit) by the year 2100. We need to keep that natural carbon in the ground, but through climate change, we are warming the planet and releasing that stored CO2, further compounding our problem. We are now facing an exponential impact caused by a chain of events that needs to be slowed and, ideally, broken.

---

[126] Nasa Earth Observatory, 'Effects of Changing the Carbon Cycle', 2011

Climate change and the continuing loss of biodiversity and degradation of ecosystems weaken their capacity to support vital recovery processes to the point where we risk reaching irreversible 'tipping points'. In the next section, I'll cover the critical role of the Earth's natural environments, how we have placed them in great danger, and what this means for our futures if we don't stop and take this opportunity to change our behaviours.

Our natural environments

The world is already experiencing the effects of climate change, including more frequent and severe storms, floods, droughts, and wildfires, which endanger our cities, communities, crops, water, and wildlife. It poses a fundamental threat to nature, species, and people. We are in the midst of a climate and nature emergency, and the two are inextricably linked. Climate change is causing the state of nature to deteriorate, and the loss of wildlife and wild areas leaves us ill-equipped to reduce carbon emissions and adapt to that change. This is urgent, and both must be resolved simultaneously.

When healthy, our natural habitats can reduce the risk of flooding, help prevent coastal erosion, enhance people's health and well-being, and maintain healthy soils, clean water, and naturally pollinate crops, thereby helping to sustain us. The International Union for the Conservation of Nature (IUCN) identifies four areas in nature where climate change will pose the greatest threat and impact. We must all gain a deeper understanding of the impact to these natural environments as they are so essential to our own livelihoods.

Soil and drylands[127]

The largest terrestrial carbon sink is soil. More carbon is stored in the world's soils than in the planet's biomass (living organisms) and atmosphere

[127] IUCN, 'Land degradation and climate change', Nov. 2015

combined. This includes soil organic carbon, which is essentially biodiversity: microbes, fungi, and invertebrates, as well as root matter and decomposing vegetation. In addition to offsetting greenhouse gas emissions, increasing soil carbon stocks through appropriate land management has other benefits for people and society.

Soil degradation is the process by which soil quality deteriorates and its capacity to support animals and vegetation decreases. Certain physical, chemical, or biological characteristics of soil that support life begin to deteriorate. Soil degradation includes soil erosion, which occurs when topsoil and nutrients are lost either naturally, such as through wind erosion, or through human actions, such as poor land management (e.g., through extensive agriculture or urbanisation in which land is overused or abused).

Loss of soil organic carbon is one of the biggest signs of land degradation, and land degradation is one of the greatest challenges to sustainable development, conservation of biodiversity, and mitigation and adaptation to climate change. According to National Geographic, more than 75 percent of the Earth's land is already degraded[128], and more than 90 percent could become degraded by 2050.

When land is degraded, soil carbon and nitrous oxide can be released into the atmosphere, making land degradation one of the most significant contributors to climate change. Land degradation has been responsible for the loss of an estimated two-thirds of all terrestrial carbon deposits from soils and vegetation since the 19th century.

Agriculture, forests, and other land-use sectors account for approximately a quarter of all human-caused greenhouse gas emissions. Sustainable land management is needed and must be accelerated through policy and financial mechanisms in order to increase soil organic carbon in a manner that combats desertification, prevents biodiversity loss, and aids in climate change mitigation and adaptation.

---

128 National Geographic, 'Desertification, explained' C. Nunez, May 2019

Oceans[129]

Increasing carbon dioxide and other greenhouse gas emissions from human activities are causing changes in water temperature, ocean acidification, and deoxygenation, leading to changes in oceanic circulation and chemistry, rising sea levels, extreme weather, and changes in the abundance and diversity of marine species.

Climate change weakens the capacity of the ocean to provide critical ecosystem outputs such as food and oxygen, as well as to support nature-based solutions to climate change adaptation. The degradation of coastal and marine ecosystems endangers the physical, economic, and food security of local communities.

The ocean serves a vital role in regulating the climate of the planet. According to the IPCC[130], the ocean has absorbed approximately 30 percent of manmade carbon dioxide from the atmosphere and 93 percent of the extra energy resulting from the enhanced greenhouse effect. However, this has resulted in increased ocean stratification (the prevention of water mixing due to the different properties of distinct water masses), changes in ocean currents, and the expansion of oxygen-depleted zones.

Currently, changes in the geographical ranges and growing seasons of marine species, as well as the diversity and abundance of species communities, are being observed. At the same time, weather patterns are altering and the frequency of extreme events is rising. The warming of the atmosphere is causing inland glaciers and ice to dissolve, resulting in rising sea levels that have significant effects on shorelines (coastal erosion, saltwater intrusion, habitat loss) and coastal human settlements. The same IPCC report projects that the global mean sea level will rise by between 0.43

---

[129] IUCN, 'The Ocean and climate change', Nov. 2017

[130] IPCC, WGIIAR5-Chapter 30 'The Ocean', Hoegh-Guldberg, O., R. Cai, et al

and 0.84 metres by 2100 relative to 1986-2005 figures. It is anticipated that the frequency of El Nino events will increase due to rising GHG emissions.

Additionally, CO2 emissions are acidifying the ocean, making many marine species and ecosystems more vulnerable. As a result of ocean acidification, marine organisms such as corals, plankton, and crustaceans are less able to construct their shells and skeletal structures. Moreover, it exacerbates existing physiological stresses (such as impaired respiration and reproduction) and reduces the growth and survival rates of some species during their early life phases.

"The ocean is not just 'the lungs of the planet' but also its largest 'carbon sink' – a vital buffer against the impacts of climate change."

United Nations[131]

In the race to tackle climate and nature emergencies, the ocean is our friend. It absorbs over 90 percent of heat and almost a third of all the carbon dioxide that humans have ever produced. It even provides the air we breathe; every second breath we take comes as a result of the ocean. Our actions are damaging Mother Nature's greatest gift to our very existence.

Species[132]

The 1°C increase in global mean temperature has severe and often unexpected effects on species, affecting their abundance, genetic composition, behaviour, and ability to survive. Species extinctions threaten the processes that nature provides for humans, such as acting as carbon sinks and enhancing our climate change resilience.

---

[131] United Nations, 'The Ocean - the world's greatest ally against climate change', (undated)

[132] IUCN, 'Species and climate changes', Oct. 2021

Human-caused climate change is already impacting species, and its rapid onset is limiting the ability of many species to adapt to their environments. At least 42,100 species on the IUCN Red List of Threatened Species™[133] are currently threatened by climate change, increasing the likelihood of their extinction.

In addition to increased disease rates and degraded habitats, climate change is causing species-specific changes that threaten their survival. Species are the building blocks of ecosystems, and reducing the threats that climate change poses to species is absolutely critical. Therefore, helping species adapt to altering climates and prepare for extreme weather events such as heat waves, floods, and droughts will require us to create far more effective conservation responses.

Environmental policies aimed at reducing carbon dioxide emissions are crucial for mitigating the effects of climate change on species, and prioritising nature conservation and adopting strategies to promote climate change adaptation will enhance species survival. According to conservation organisations in the UK that have been studying decades of research, and as reported recently by the BBC[134], we are failing in our response:

"The State of the Nature report found 16 percent of 10,000 mammals, plants, insects, birds and amphibians assessed were threatened. Plants and animals monitored since the 1970s have declined in abundance by an average 19 percent."

BBC, Sept. 2023

To put this into perspective, the diversity and abundance of natural fauna in this country have decreased by 19 percent since my arrival in the 1970s. As a direct result of human activity and climate change, there are now one-fifth

---

133 IUCN 'Red List of Threatened Species™', see https://www.iucnredlist.org

134 BBC, 'Nature crisis: One in six species at risk of extinction in Great Britain', V. Gill and K. Stephens, Sept. 2023

fewer native plants and animals when I take Belle for a walk in the Suffolk countryside. The effect we are having is devastating and real.

Forests and natural land types[135]

Forests and natural land types contribute to climate stability. They regulate ecosystems, safeguard biodiversity, play a crucial role in the carbon cycle, sustain livelihoods, and contribute to sustainable growth. To optimise their climate benefits, we must preserve more natural land types, manage them in a more sustainable manner, and restore more of those that have been lost.

According to the IUCN, halting the loss and degradation of natural systems and promoting their restoration have the potential to contribute over one-third of the total climate change mitigation scientists say is required by 2030. Forests' function in climate change is two-fold. They are both the cause and a solution for greenhouse gas emissions. According to the IUCN, approximately 25 percent of global emissions originate from the land sector, making it the second-largest source of greenhouse gas emissions after the energy industry. Approximately fifty percent of these (5-10 GtCO2e annually) are caused by deforestation and forest degradation.

Forests are also one of the most essential means of mitigating climate change's effects. Every year, forests absorb approximately 2.6 billion metric tonnes of carbon dioxide, or one-third of the CO2 produced by the combustion of fossil fuels. In addition, they estimate that nearly two billion hectares of degraded land across the world, an area larger than Russia, present restoration opportunities. Increasing and sustaining natural forests can therefore be a major climate change solution.

---

135 IUCN, 'Forests and climate change', Feb. 2021

But it is not all about forests. The Wildlife Trust[136] in the United Kingdom cites the following examples of non-forest natural land types that highlight the importance of natural land diversity for climate mitigation:

Peatland

The peatland soils of the United Kingdom contain approximately 3.2 billion metric tonnes of carbon but are severely degraded and emit the equivalent of 23 million metric tonnes of CO2 annually. Restoring them is one of the most cost-effective, nature-based solutions for preventing this emission.

Grassland

The grasslands of the United Kingdom contain 2 billion metric tonnes of carbon, but this carbon is vulnerable to disturbance. From 1990 to 2006, the conversion of grasslands to arable land released 14 million metric tonnes of carbon dioxide. Restoring species-diverse grasslands to sequester carbon and sustain an abundance of fauna should be a priority in the UK.

Woodland

About one billion metric tonnes of carbon are sequestered in UK woodlands, primarily stored in woodland soil and sediments. This must be carefully managed to maximise benefits and avoid harming other habitats.

Wetlands

Wetlands are able to store carbon for centuries, but in some regions of the United Kingdom, over 90 percent of the wetland habitat has been lost. Restored wetland provides abundant natural habitats for species, naturally cleans water, and reduces downstream flood risk.

Stopping the loss and degradation of natural ecosystems and promoting their restoration have additional benefits. A quarter of the world's population, or 1.6 billion people, depend on these natural lands for their

---

136 Go to https://www.wildlifetrusts.org/ for more information

livelihoods, including many of the world's poorest nations. According to the UN Environment Programme[137], forests alone contain "80 percent of amphibian species, 75 percent of bird species, and 68 percent of the world's mammal species. Forests are home to 80 percent of the world's terrestrial biodiversity."

Supporting nature, particularly through the restoration and maintenance of forests and natural ecosystems, can and must be a part of the solution for limiting global warming.

Summary

> **"Nature-based solutions must be combined with slashing our global greenhouse gas emissions. Taking a more holistic approach by tackling the climate and biodiversity crisis together will benefit both and offers us the best chance for a future where the planet can support us."**
>
> Natural History Museum[138]

Climate change damages biodiversity. It is one of the main causes of biodiversity loss. At the same time, climate change will accelerate further if biodiversity and ecosystems are not effectively protected. By conserving nature and restoring ecosystems, we reduce vulnerability and increase resilience. Nature conservation and restoration are major, cost-efficient allies in our fight against climate change.

---

137 UN Environment Programme, 'Earth's biodiversity depends on the world's forests', May 2020

138 nhm.ac.uk, 'How are climate change and biodiversity loss linked?', Chetan-Walsh and Hendry, (undated)

As examples of how we can benefit from nature's ability to keep us safe, the European Commission[139] identifies the following:

- By using appropriate agricultural and forestry practices to increase water retention capacity, we can mitigate the occurrence of droughts.
- By maintaining and restoring wetlands and riverbeds, we can better control the impact of river flooding.
- By cultivating diverse forests, we can reduce the risk of wildfires starting.

I have reread this section many times. The figures revealing the damage we are causing are shocking and depressing, and each time I put down the book, I am unable to fathom why and how we have let things get this bad. The fact is that Mother Nature possesses a scaled healing power that humans cannot replicate, but we do have the ability to mistreat and damage her, which can lead to a dramatic decline in her ability to help us live in a safe and stable world.

As technologists, we can play our part in preserving nature's ability to maintain balance. We must keep in mind that every kilogramme of carbon dioxide saved in the creation and operation of our digital products and services contributes towards stopping us from reaching irreversible tipping points and an exponential decline in nature's capacity to support us. So the remainder of this book is a look forward, a build on my recipe for change, and an attempt to help IT leaders drive our sector to the forefront of decarbonisation.

At the beginning of the book, I suggested it was a choice to take action. I've changed my mind. We must.

---

139 Reference: European Commission, 'Nature's role in climate change', Aug. 2009

The AI Balancing Act

This section will compare the environmental impact of artificial intelligence with its role as a driver and accelerator for sustainability. When the headwinds propelling the increased use and adoption of AI in society and business are so strong, it will require strong leadership to manage this balance.

"Artificial intelligence (AI) technology can help us fight climate change – but it also comes at a cost to the planet. To truly benefit from the technology's climate solutions, we also need a better understanding of AI's growing carbon footprint."

Horizon, The EU Research & Innovation Magazine[140]

The dilemma

Earlier in the book, I talked about the carbon footprint caused by the energy consumption of Generative AI, using ChatGPT-4 as an example. It should come as no surprise that large language and foundation models that process a huge variety and volume of data to translate, predict, and generate content have significant computational demands and that this needs more data centre resources, resulting in more energy usage and an escalating environmental impact.

However, AI can also advance the digital transformation of our economies and societies, scaling our cognitive abilities, enhancing healthcare, transportation, and education, creating improved customer-centric experiences, and having a profound impact and potential to accelerate

[140] Horizon, EU Research and Innovation, 'AI can help us fight climate change. But it has an energy problem, too', Sept. 2019

human efforts. This offers a genuine opportunity and glimmer of optimism for addressing our greatest challenge: reaching Net-Zero.

According to a study by PWC[141], "AI levers could reduce worldwide greenhouse gas (GHG) emissions by 4 percent in 2030, an amount equivalent to 2.4 Gt CO2e–equivalent to the 2030 annual emissions of Australia, Canada, and Japan combined." I'm not sure how these numbers were calculated, but the idea that AI can help us optimise everything we do, every process, and every production and ultimately contribute towards a more sustainable world is believable. And therein lies the dilemma, the balancing act that IT leaders must help address: making conscious decisions to use AI for good in a sustainable manner while recognising and managing its environmental impact.

A study

In 2020, Nature.com[142] completed a study that examined both the positive and negative effects of AI on the United Nations' Sustainable Development Goals (SDGs). The study recognised how artificial intelligence could either help or hinder the delivery of all 17 goals (and the 169 targets set forth in the UN's 2030 Agenda for Sustainable Development).

The findings of their study are summarised in the diagram on the next page, which shows AI's potential to act as either an enabler or an inhibitor for each of the SDGs. The numerals within the coloured squares represent each of the Sustainable Development Goals. The percentages on top represent the proportion of all targets that could be affected by AI, while the percentages in the inner circle correspond to the proportions within each SDG. The

141 PWC report, 'How AI can enable a sustainable future' (undated)

142 Nature Communications, 'The role of artificial intelligence in achieving the Sustainable Development Goals', Jan. 2020

diagram's outer circle also displays the results corresponding to the diagram's three major groups: society, economy, and environment.

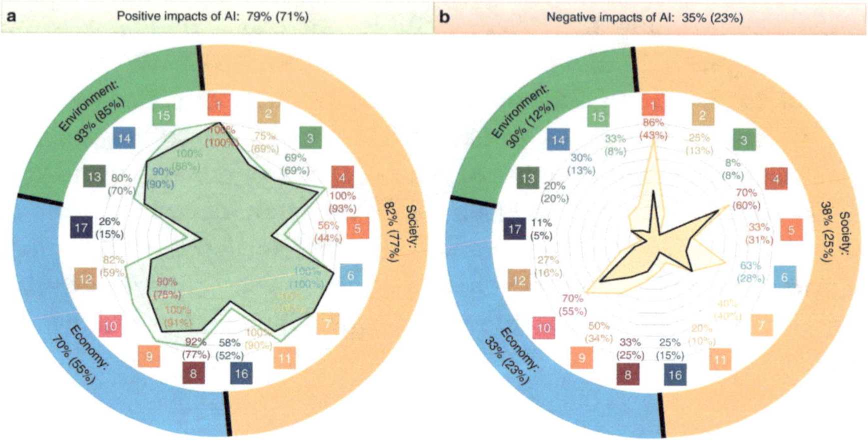

Positive impacts accounted for 79 percent of all SDGs, while negative impacts accounted for 35 percent. When environmental SDGs were prioritised, the results were even more beneficial, with a 93 percent positive impact.

AI and environmental outcomes

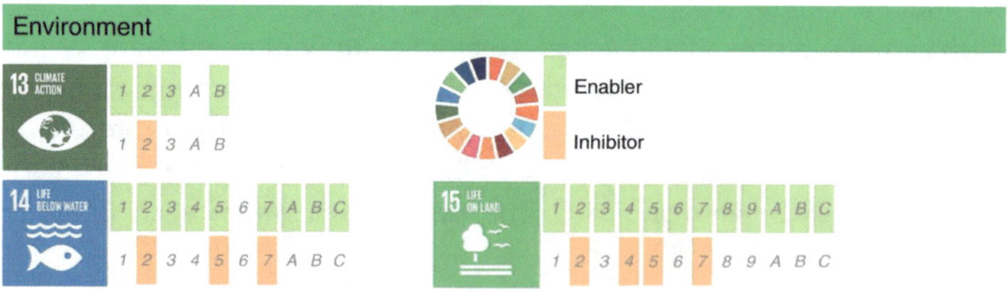

The three SDGs in the 'Environment' group (SDGs 13, 14, and 15) are related to climate action, life below water, and life on land. For the Environment group, 25 goals (93 percent) for which AI could serve as an enabler were identified; they are summarised as follows:

- The benefits of artificial intelligence are derived from analysing large-scale, interconnected databases in order to develop cooperative environmental preservation measures.
- The development of AI will facilitate understanding climate change and the modelling of its possible impacts.
- AI supports low-carbon energy systems with high renewable energy integration and energy efficiency, all of which are required to combat climate change.
- AI can also be used to improve the health of ecosystems, for example, to prevent and substantially reduce all types of marine pollution by automatically identifying potential oil spills using AI algorithms.
- Using neural networks and object-oriented techniques to enhance the classification of vegetation cover categories from satellite images in order to combat desertification and restore degraded land and soil.
- Identifying the main drivers of desertification through the use of AI to identify desertification trends over large areas, thereby providing information for environmental planning, decision-making, and management to prevent further desertification or to help reverse trends.

A really strong set of positive enablers, but the study also concluded a need for balance, stating:

> **"Efforts to achieve SDG 13 on climate action could be undermined by the high-energy needs for AI applications."**

A positive outlook

We must use technology to do good and accelerate the missions that matter. The outlook is positive, but only if we make informed choices and decisions and constantly balance the benefit against the environmental cost of using more technology, as well as more advanced technology.

> **"AI can play a role in tackling environmental challenges, from designing more energy-efficient buildings to monitoring deforestation to optimising renewable energy deployment. This can be on a large scale – such as satellite monitoring of global emissions, or a more granular scale – such as a smart house automatically turning off lights or heat after a certain time."**
>
> United Nations Environment Program[143]

AI can support decarbonisation across all four major industrial sectors: agriculture, water, energy, and transportation. In these four sectors, environmental AI-based applications have the potential to contribute to mitigating Earth's environmental challenges. According to the PWC report, these four sectors represent around 60 percent of GHG emissions and are critical to our environmental and earth systems, for example:

- AI can assist in transforming agricultural production by enhancing the monitoring and management of environmental conditions and crop yields, which are required to support global population growth and the resulting demand for food.
- Through intelligent grid systems, AI has the potential to increase energy sector efficiency by managing demand and supply and optimising renewable energy solutions.
- Through more precise autonomous vehicle technologies, traffic prediction, and real-time journey planning, AI applications in the transportation sector can facilitate the safe, efficient, and environmentally friendly movement of cargo and people.

---

143 United Nations Environment Program, Digital Transformation sub-programme, 'How artificial intelligence is helping tackle environmental challenges', Nov. 2022

- Utilising AI for water resource management, forecasting, and monitoring can alleviate the global water crisis by reducing and eliminating waste, lowering costs, and minimising environmental impacts.

So in addition to the work required to determine data ownership, the ethical use of AI, building a trust framework, appropriate regulation, and reaching consensus on when to use AI or not, continuous measurement of AI's environmental impact is essential. AI will only become a truly positive force for accelerating sustainability and our mission to reach Net-Zero when all of these are in place.

The Case for Sustainable IT

"The greatest opportunity for businesses is that once an organisation has a sustainable IT estate, this can be used to unlock greater sustainability outcomes"

Matt Bradley, Capgemini

Sustainable IT can help businesses create a more resilient organisation by mitigating ESG risks, improving their reputation, and enhancing their standing with consumers, employees, and business partners. In addition, transitioning to energy-efficient solutions can result in potential cost savings and boost financial performance.

One of the biggest obstacles for organisations attempting to implement sustainable IT practices is the need for a measured baseline and a greater understanding of their digital environmental impact. This is a very real challenge, as, according to the Capgemini Research Institute, 43 percent of technology executives are unaware of the size of their organisation's IT-related carbon footprint[144]. I actually think this number is conservative based on my interactions with IT leaders across many organisations.

Action needs to be taken to establish a context and a case for change. This will help identify solutions that will have the greatest material impact on sustainability. Identifying metrics to monitor the impact of sustainability-based investments and supporting upcoming reporting regulations that will cover the entire IT estate are key activities that need to be part of those solutions. Leaders in technology also have to contend with sustainability-related data that may not always be complete or perfect, so their decision-making and prioritisation will need to be dynamic and flexible. As a result of these challenges, the need for strong leadership in technology to help drive

---

144 capgemini.com, 'Sustainable IT is the backbone of a greener future' and 'Sustainable IT: Why it's time for a Green Revolution for your organisation's IT' (undated)

immediate action is now more important than ever. I think the following quote by Maria Mendiluce, CEO of the We Mean Business Coalition, sums up what is needed perfectly:

> **"We can't afford to let perfection be the enemy of taking climate action. I would rather see a million companies worldwide roll up their sleeves and getting on with delivery of imperfect climate plans today, rather than lose another decade trying to be faultless. Those companies pioneering emissions reductions today are paving the way for others to follow."**
>
> We Mean Business Coalition[145]

The Case for Action

The main cause of climate change and the environmental issues facing us all today is overconsumption, so finding ways to lower consumption across all sectors, including ICT, is critical. Reduction and efficiency have to be the fastest ways of making a difference and will create the capacity needed to support the continued growth of digital. If we don't do this as a sector, then we are part of the problem, and every decision to defer or not take action makes us part of that problem.

The Intergovernmental Panel on Climate Change (IPCC), formed by the top climate scientists in the world, published the sixth assessment report's (AR6) fourth and final instalment in March 2023. The so-called synthesis report is the culmination of an eight-year effort by the world's most authoritative scientific body on climate change. This included the findings of 234 scientists on the physical science of climate change, 270 scientists on impacts, adaptation, and vulnerability to climate change, and 278 scientists on climate

[145] Maria Mendiluce, CEO, We Mean Business Coalition, 'We can't afford to let perfection be the enemy of companies taking climate action', Feb. 2023

change mitigation. The IPCC synthesis report provides the most comprehensive and best available scientific assessment of climate change. Having studied it after publication, my view is that it should be compulsory reading for everyone.

Evidence of global warming already underway

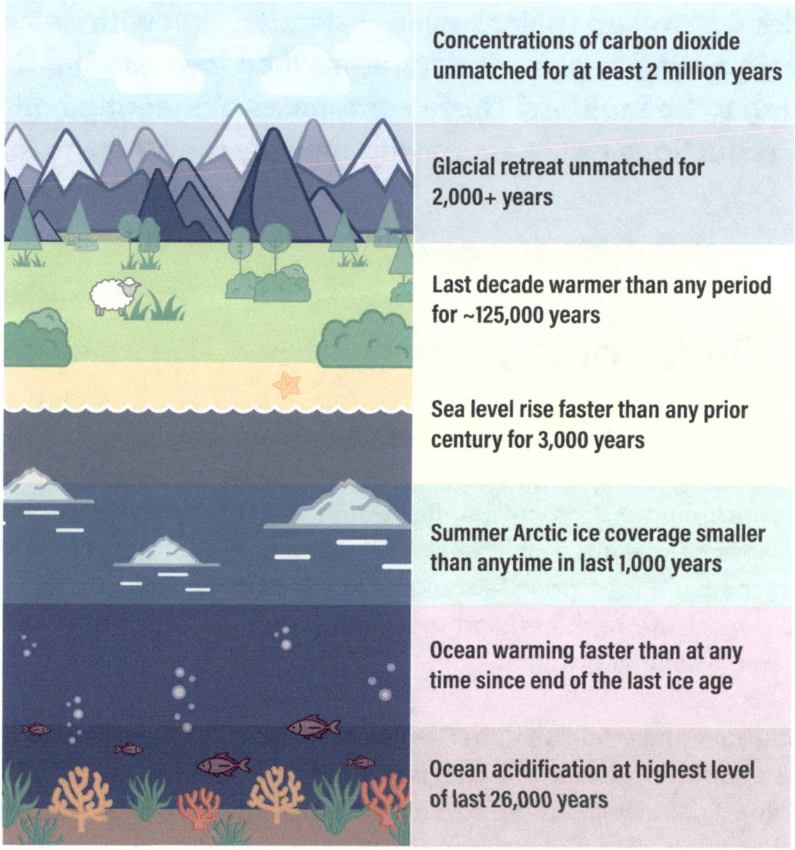

Source: IPCC AR6.

WORLD RESOURCES INSTITUTE

The diagram above, taken from the report, summarises the impact we have already inflicted on the planet. The full report makes for harder and even more compelling reading. Over nearly 8,000 pages, the AR6 details the detrimental effects of rising greenhouse gas (GHG) emissions worldwide, as well as the increasingly dangerous and irreversible risks if we do not alter our course.

Despite this destruction, the IPCC continues to offer us hope by highlighting pathways to avoid intensifying these climate risks. It identifies readily available and, in some instances, highly cost-effective actions that can be taken immediately to reduce greenhouse gas emissions, scale up carbon removal, and strengthen resilience. While the window of opportunity to address the climate crisis is rapidly closing, the IPCC believes that a safe, habitable future is still achievable.

This is called positive optimism, and for now, I wholeheartedly subscribe to it. However, I have placed my own time window on how long this will last. If all the talking and protests don't translate into real, decisive action over the next 18 months, then I think we enter a time where, frankly, I think we will have missed our opportunity to make a difference. It will be a generational failure, and we can't let that happen.

The AR6 report makes clear that the risks of inaction are enormous and that the path forward requires unprecedented levels of change. The report also serves as a reminder that we have never had more information regarding the severity of the climate emergency and its cascading effects, as well as the steps that must be taken to mitigate the intensifying risks. In its own words, there is no room for denial and no rationale for inaction.

"A narrow window of opportunity is still open, but there's not one second to waste."

According to the findings of the report, it is still possible to limit the global temperature rise to 1.5°C (2.7 degrees Fahrenheit), but only if we act now. The IPCC makes it plain that the world must peak GHG emissions by 2025 at the latest, nearly halve GHG emissions by 2030, and reach Net-Zero CO2 emissions by the middle of the century, while also ensuring a just and equitable transition. Adapting to this new world requires an 'all-hands-on-deck' approach to ensure that communities suffering from the worsening effects of the climate crisis have the means to do so. Governments, the private sector, civil society, and individuals must all take action to ensure that we achieve the future that we want.

The mirror has been held up. The time has come for all technology leaders to stop observing the problem and prioritise taking action. Please use your influence to positively make a difference.

How much can we achieve?

It is time to prioritise reducing energy consumption and increasing energy efficiency across everything we do.

> **"The only way to keep humanity safe is the immediate and sustained radical cuts to greenhouse gas emissions. If we want to keep people safe then large and sustained cuts to carbon emissions need to happen now. That is the very simple acid test that must be applied to all climate policies. The time for wishful thinking is over."**
>
> The Conversation[146]

According to the International Energy Agency (IEA)[147], simultaneously reducing consumption and increasing energy efficiency can provide substantial environmental benefits by reducing greenhouse gas (GHG) emissions, both directly from fossil fuel combustion or consumption and indirectly from electricity production. Energy efficiency is one of the most

---

146 The Conversation, 'Climate scientists: concept of net zero is a dangerous trap', James Dyke (University of Exeter), Robert Watson (University of East Anglia) and Wolfgang Knorr (Lund University), April 2021

147 IEA, 'Energy Efficiency 2021', Nov. 2021

important means by which the world can fulfil its energy needs with less energy consumption, which is central to the majority of the IPCC's GHG emission pathways for limiting global warming to 1.5°C.

Efficient World Scenario

Based on IEA modelling developed as part of Energy Efficiency 2018[148], if the world had implemented all cost-effective energy efficiency measures based on existing technology, it would have led to a peak in energy-related greenhouse gas emissions before 2020. According to this projection, outlined in the IEA's Efficient World Scenario (EWS), energy efficiency would reduce annual energy-related emissions by 3.5 Gt CO2-eq (12 percent) compared to 2017 levels, delivering more than 40 percent of the reduction needed to be in accordance with the Paris Agreement.

In conjunction with renewable energy and other reduction measures, energy efficiency is essential to achieving global climate objectives. The IEA's graphs show greenhouse gas (GHG) emissions from 2000 to 2017 (left) and from 2000 to 2040 (right), with and without efficiency in the energy sector. They are based on the New Policy Scenario (NPS), which is structured on a new global context set by the 2009 Copenhagen Accord, which includes all national pledges and announced policy intentions, and the EWS.

148 IEA, 'Energy Efficiency 2018, Analysis and outlooks to 2040', Oct. 2018

So why aren't we travelling on the downward green line? Sadly, the EWS has remained a hypothetical scenario. Instead of reducing consumption and increasing efficiency, we have increased our energy consumption, and the resulting carbon emissions have brought us to a critical point. The growth of the ICT industry and the expansion of digital have contributed to this increased usage.

As stated in the IPCC Synthesis Report's call to action, there is still hope, but it is now up to each of us to play our part, and the wonderful thing about digital and ICT is that we already have the tools we need to begin making a difference.

So let's start.

... but before you do ...

Before you move on to the next section, take a moment to refresh your mind on my 'recipe for change' and the Digital Decarbonisation Framework on page 79. Then revisit the 'mishmash' architecture diagram on page 265, and remind yourself of the complexity of your starting position and the different solutions you may need to use to move from left to right. Finally, look back at your screening results and your reflections gained from page 259 to consider the scale of opportunity that is in front of you. There's a lot to contend with, right? The next sections are here to help you put everything together into a sustainable IT strategy that you can effectively execute.

Building a Successful Sustainable IT Strategy

Let's start.

The following diagram covers the best practice steps needed to build a foundation for developing a sustainable IT strategy. This approach is most effective when fully aligned with your organisation's global sustainability strategy.

4 Best Practice Steps to Building a Sustainable IT Strategy

1. **Tool Selection**
   Select tools that can help generate the data and insight across own emissions and in your supply chain to be able to inform priorities and decision making

2. **Measurement and Materiality**
   Measure the environmental footprint of IT and use the data to better understand the solutions that offer the most impact

3. **Governance**
   Create effective governance procedures that support the sustainable IT strategy, including setting & tracking targets, KPIs or OKRs and investment decision making criteria based on sustainability

4. **Operationalise**
   Execute the plan by embedding sustainability into technical architecture decisions; software design, build and operation; procurement and people & culture. Utilise the People, Process, Technology (PPT) framework as a basis for driving change.

✓ **Sustainable IT Strategy**
Create a sustainable IT strategy that aligns with the organisations global sustainability strategy

Imagine, as a technology leader, being able to fully understand the carbon impact of your IT estate. Imagine empowering your teams with the data necessary to make energy efficiency changes at the software application level. Imagine being able to reduce the environmental impact of digital products while achieving both operational and cost efficiencies.

To achieve all this, technology leaders should follow the four best practices steps in the preceding diagram when defining and developing their sustainable IT strategies. One of the best first things an IT leader can do when starting is to evaluate the IT infrastructure and software they already own. Get this context as quickly as possible through screening your technology landscape, as it will form your baseline position and a place from which you can take informed action. Once you have decided to create a baseline, quickly establish which measurement tools are best suited to performing the initial screening, and then subsequently provide the detailed measurements necessary to identify and prioritise actions that offer the most effective carbon reduction and contribution to your organisation's decarbonisation commitments.

Governance can hinder or accelerate action, get it right and you help provide clear goals, targets, and timelines that can be widely syndicated and communicated. It should define policies that mandate consistent reporting of the enterprise-wide IT carbon footprint and enable technology leaders to effectively operationalise their strategy by, for example, integrating sustainability into technical architecture and software decisions, through your supply chain decisioning and procurement processes, and by fostering a sustainability-focused culture among employees.

"Pace, not perfection is the key to reducing emissions."

National Infrastructure Commission[149]

---

[149] James Heath, CEO National Infrastructure Commission, July 2023

Later in this chapter, I'll discuss two topics that all IT leaders will want to better understand: (1) the impact of new reporting standards and (2) how to construct an investment case to mobilise your sustainable IT strategy. But ahead of that, let's expand on the best practices, remembering that pace, not perfection, is required. I hope that by the end, you will have everything you need to mobilise quickly and mitigate any early risks associated with decarbonising digital in your organisation.

1. Tool selection

Until recently, there were very few tools available to assist technology leaders in understanding the climate impact of their IT. The marketplace for tools is becoming increasingly sophisticated, with greater breadth and coverage, although some IT leaders have recently said to me that they feel it is now becoming overcrowded and confusing. I consider these tools to be part of a larger ecosystem of interlocking measurement capabilities. There is no single tool that does everything, and the skill is to know when to use the right tool for the right job.

My view is that we need tools to be able to achieve different but interconnected outcomes; for example, establishing a baseline and assisting in setting reduction targets are very different from workload monitoring or source code performance analytics. I've mapped these outcomes to a hierarchy that I think works, as shown in the diagram below.

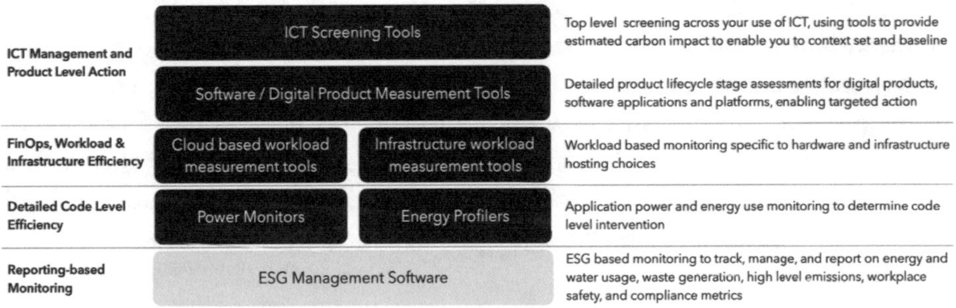

Remember that your delivery maturity and the scope of your IT-related decarbonisation targets will determine which tool or combination of tools should be used to support action at any given time.

ICT Screening Tools

These tools have been designed to provide a rapid assessment of the estimated energy and carbon impact of an organisation's use of the main ICT components (i.e., desktop managed services, networks and telecoms, and cloud and data centre services). They can help technology leaders establish a baseline for their entire IT infrastructure estate using a standardised and consistent calculation methodology across each hardware category of ICT while managing the inconsistency of coverage and the quality of manufacturers primary data.

Software / Digital Product Measurement Tools

These are digital product or software application-specific measurement tools that cover the main lifecycle phases of a software product. These assessments are activity-based at the digital product level and can provide a baseline carbon emissions value, product ratings, carbon intensity scoring, and targeted life-cycle stage actions to reduce carbon impact.

Cloud and Infrastructure-based Workload Measurement Tools

These cover open-source cloud carbon measurement tools (such as Cloud Carbon Footprint), manufacturer- or CSP-specific tools (e.g., IBM Turbonomics, Amazon CloudWatch, Microsoft System Centre), and independent cloud management products that have added sustainability-related components (e.g., Dynatrace). The use of these tools requires careful oversight to ensure consistency of approach and alignment of calculation results. Also, consider that some use spend as a proxy rather than actual usage to drive calculations. Finally, take into account that real-time, workload-based monitoring will in itself incur significant energy usage overheads, so it is important to use them when targeting next-level action at the engineering level.

Some organisations will have access to energy consumption at the infrastructure level in their data centres. This can be converted into carbon emissions using the correct emission factors. If only this data is used, then other important factors, such as embodied carbon and operational activities associated with the infrastructure, will need to be calculated separately.

Power Monitors and Energy Profilers

Power monitors are hardware tools that analyse the actual power consumption of a device or component at any given time by connecting to its power source. Energy profilers, on the other hand, are typically based on an estimation model of the power cost of the various hardware components and do not require specialised hardware or power sensors. These tools are useful when code-level energy usage data is needed. Note that they are expensive to set up and maintain and can consume significant amounts of energy. They should only be used when targeting performance in specific applications or components of underperforming applications, with expected benefits known to outweigh the cost of operation.

ESG Management Software

These products collect data and allow businesses to monitor, manage, and report on energy and water consumption, waste generation, greenhouse gas emissions, workplace safety, and compliance metrics. It is anticipated that, over time, these reporting tools may begin to incorporate IT-level activity-based metrics via APIs to specialist measurement platforms and provide an organisation with a complete view of its emissions impact.

The best method to achieve an accurate and comprehensive understanding of the energy and carbon impacts of an IT estate is, at the top of the hierarchy, to select independently developed tools that are not aligned with manufacturers or

service providers. As you drop down the layers, service provider alignment can provide improved integration and ease of data acquisition. By deploying these tools and implementing best practices, businesses can improve the integrity of their IT-related sustainability efforts.

It should come as no surprise that the tools I've developed at GoCodeGreen can help you as you establish your sustainable IT strategy. If you look again at the diagram on page 309, the measurement tools that GoCodeGreen offers cover the 'ICT Management and Product Level Action layers'.

2. Measurement and materiality

I referred to a 'hierarchy' of measurement tooling in the previous section, which covered the 'how' in terms of measurement. Next, we need a best practice for determining 'what' to measure and in which order. Technology leaders should be focused on establishing a context for measurement and deciding in what sequence to approach it based on the materiality of the impact. In my experience, first establishing a high-level context before moving on to product-level specifics is the best and most effective strategy. This can be quickly followed by undertaking usage analysis, power monitoring, and code-level energy analysis when it is required; all of this enables both a broad and deep decarbonisation of digital

Compare the sequencing strategy described above to the analogy for prioritisation based on fitting large rocks, pebbles, and sand into an empty jar. If you begin filling the jar with sand and then pebbles, there will be no space for rocks. A better approach is to begin with the large rocks, which represent the things that are most important and will have the greatest impact, then the pebbles, which represent everything of medium importance, and finally the sand, which represents all of the smaller items that, when combined, help to fill the jar. Follow the diagram below from top to bottom, as indicated by the green arrow:

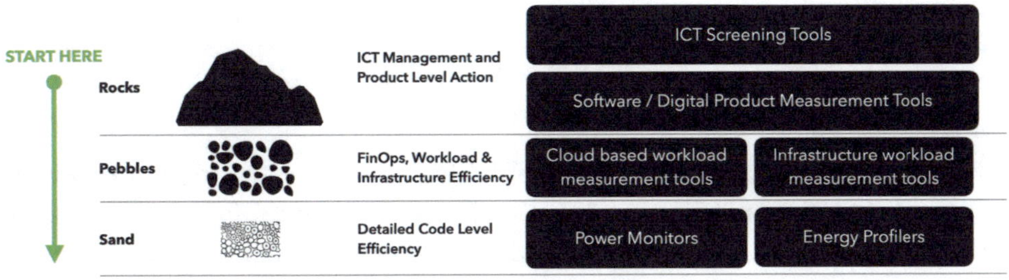

As a result of measuring in this sequence, GoCodeGreen has identified material and significant opportunities to enhance IT's energy efficiency and reduce its carbon footprint. What follows are examples based on assessments completed with real clients. Here, we would start by first examining the IT estate's 'big rocks', or the major technology groupings according to the GHG ICT Sector Guidance. This would create the organisational context for taking action, for example:

	Total Estimated Annual Carbon Impact (tCO2e)	Total Estimated Annual Energy Usage (kWh)
Desktop Managed Services	237	365,227
Telecoms & Network Managed Services	268	763,044
Cloud & Data Centre Services - On Premise	648	1,034,016
Cloud & Data Centre Services - AWS	134	160,685
Cloud & Data Centre Services - Azure	318	383,040
Total	1,604.73	2,706,012

Next, we would consider a selection of the major software products that utilised the hardware components identified in the screening phase. By calculating the carbon impact at life cycle stages, such as the two phases of software release and use as illustrated in the table on the next page, it was possible to identify targeted actions for the client's teams to focus on.

	Lifecycle Phase	Total Estimated Annual Carbon Impact (tCO2e)
Digital Products	Release	132
	Use	296
Total		428

After this, we would move into 'pebble' territory. Very few digital products will have identical profiles, and each will have a unique impact and set of reduction actions based on its lifecycle stage and its own history of choices during design and development. At GoCodeGreen, we consistently identified double-digit opportunities to reduce the carbon footprint of digital products and services, including their underlying technology value chain. The table below shows the different actions for different products in different companies at various lifecycle stages, demonstrating the variation in the scale of carbon emission reduction possible:

UK Financial Services
Internal, modern, design stage assessment
Stage: Core (pre-build)
Carbon reduction opportunity **17%**

ESTIMATED REDUCTION OPPORTUNITIES:	tCO2e	%
Compute	0.66	16.2%
Demand	0.00	0.0%
Desktop Compute	0.35	8.7%
Efficiency	1.12	27.7%
External Resources	0.30	7.3%
External Travel	0.44	10.9%
Hygiene	1.07	26.3%
Internal Resources	0.00	0.0%
Internal Travel	0.16	4.0%
Location	-0.05	-1.2%
TOTAL	**4.06**	**100.0%**

Global Software Company
Part modernised, SaaS solution
Stage: Iteration (annual release)
Carbon reduction opportunity **28%**

ESTIMATED REDUCTION OPPORTUNITIES:	tCO2e	%
Compute	40.83	18.2%
Demand	0.01	0.0%
Desktop Compute	0.84	0.4%
Efficiency	89.90	40.1%
External Resources	0.44	0.2%
External Travel	0.40	0.2%
Hygiene	75.36	33.6%
Internal Resources	0.44	0.2%
Internal Travel	1.42	0.6%
Location	14.58	6.5%
TOTAL	**224.22**	**100.0%**

Global Investment Bank
Internal, on-premise data application
Stage: Use (annualised)
Carbon reduction opportunity **64%**

ESTIMATED REDUCTION OPPORTUNITIES:	tCO2e	%
Compute	12.54	24.3%
Demand	0.50	1.0%
Engineering	7.35	14.2%
External Resources	0.00	0.0%
External Travel	0.00	0.0%
Hygiene	3.57	6.9%
Internal Resources	2.65	5.1%
Internal Travel	0.54	1.1%
Location	4.26	8.3%
Service	20.20	39.1%
TOTAL	**51.61**	**100.0%**

Observing the materiality of the different ICT carbon cost categories, as shown in the table above, enables a more effective approach to decarbonisation. Once this level of measurement has been completed, it makes perfect sense to proceed to the 'sand' level to identify opportunities to reduce energy consumption and increase efficiency at a targeted and granular level. Here, teams can use workload analysis and code-level measurement to support and validate their reduction actions, as well as monitor the improvements they make.

A note of warning: data, data everywhere (and nothing adds up)

In most IT organisations, sustainability-related data is dispersed across departments, frequently in spreadsheets containing unstructured and structured data without any real alignment or context. It is on the basis of these inconsistently calculated and partial datasets that decisions and even entire carbon reduction strategies are being created. I have observed IT leaders throwing their hands in the air in despair and threatening to pause working on sustainability because they are being drowned by data that doesn't add up or give them a context for taking action.

Often, a mindset and belief set in, where all data must be available before any decisions can be made. This frequently results in businesses collecting as much data as possible–data that is not necessarily optimised, accurate, or even useful. Instead, technology leaders, supported by the right context, choice of tools, and measurement sequencing strategy, should start by narrowing their focus to collect consistently calculated data that informs and supports decision-making that leads to action.

As a guide, here are a few data-related suggestions for technology leaders to consider:

- It is preferable to collect better data and sometimes less data. This not only improves decision-making but also reduces the digital footprint associated with data collection.
- Frequently, technology leaders view the automation of data collection as a precondition for taking action. Learn to accept that not all data will be available and that you are going to need to make decisions based on the best available data.

- Accept that carbon emission calculations across ICT will change as data quality improves.
- Data automation needs the availability and ongoing upkeep of comprehensive, high-quality data inventories. Examine your organisation critically and determine if a centralised source of high-quality, well-maintained, and comprehensive data encompassing your IT estate is currently available. If this is the case, you have no reason not to take immediate action, as the tools to consume and utilise this data are available. However, if, like the majority of organisations, there is still a substantial amount of work to be done in this area, then accept it and start to use your initial measurement and assessment as the method for structuring your data, even if this requires manual effort.

Avoid the 'data everywhere' scenario. You'll recognise it when you are in a meeting where the attendees are examining various data sets and arguing about the relative veracity of each other's data instead of agreeing on how to use it to drive action. This is the point at which 'analysis paralysis' is likely to set in, and action is likely to be postponed. Avoid this situation at all costs.

3. Governance

IT governance frameworks provide organisations with a formal structure for ensuring IT investments support business objectives. Good IT governance promotes transparency and accountability in business operations, and in the context of IT decarbonisation, it can provide the guardrails to establish the organisational context needed for success.

Typically, IT governance is divided into five domains and summarised as follows:

- Value delivery, focusing on whether or not IT delivers value to the rest of the business.

- Strategic alignment, which asks questions to determine if the goals of IT and the organisation are aligned.
- Performance management focuses on how IT performance is managed and reported.
- Resource management focuses on whether or not IT resources are being managed effectively and appropriately.
- Risk management examines whether risks are identified, reported, and mitigated.

I now advise technology leaders to consider introducing a sixth dimension to encompass the sustainability of IT. Refer to the diagram below. This will ensure that appropriate focus is placed on meeting all the value, strategic alignment, performance, resource, and risk considerations required to successfully contribute to an organisation's Net-Zero commitments.

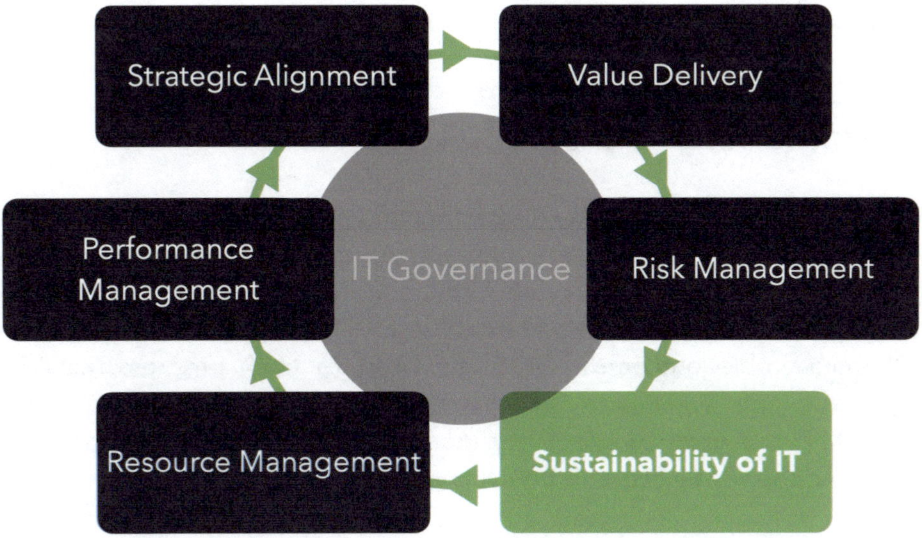

It is worth noting that all six most commonly used IT governance frameworks, i.e., ISO 3850, ISO/EIC 27000, COBIT, ITIL CMMI, and FAIR, are already being updated to reflect the importance of sustainability (e.g., ITIL 4 now clearly aligns sustainability to digital and IT). Therefore, preparing your own governance frameworks in accordance with this broader industry direction will better prepare you to meet any new standards as they emerge across the industry.

Adapting IT governance to include sustainability will help accelerate the following:

- Ensure that legal, regulatory, and compliance requirements related to sustainability can be met.
- Contribute to reducing climate risk.
- Support the organisation's Net-Zero objectives and ensure that IT-based sustainability objectives are aligned with those of the broader organisation.
- Promote climate-oriented innovation.
- Provide businesses with a competitive advantage.
- Ensure that the appropriate policies, processes, and procedures based on sustainability are consistently applied across the organisation.

Sustainability-based OKRs

Many large organisations use Objective and Key Results (OKRs) and Key Performance Indicators (KPIs) to manage and monitor delivery against their targets and goals. OKRs represent a collaborative goal-setting methodology used by teams and individuals to establish challenging, ambitious objectives with measurable outcomes. OKRs are used to track progress, establish alignment, and encourage engagement around measurable objectives. While KPIs are used to effectively measure performance at a lower level relative to specific targets and benchmarks.

OKRs with a focus on IT sustainability can provide an effective framework for defining precisely which aspect of business sustainability you intend to address. These objectives may be part of a broader business OKR#1 or a discrete technology-driven OKR#2, as indicated in the following example:

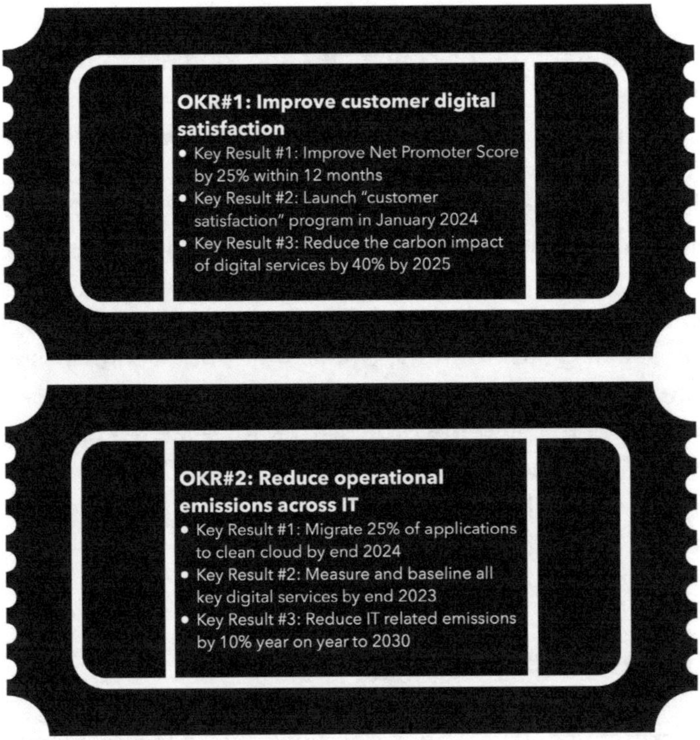

By establishing specific sustainability-related goals, organisations can set clear pathways to attaining them. This ensures that the organisation is taking the necessary measures to meet its environmental goal of reducing its carbon footprint. Imagine taking the two OKR examples above and adding them to your organisational goals. Do they make sense? I suspect that with only minor adjustments, they would work.

Alongside the benefit of having a sustainable IT goal at the organisational level to help drive action, further advantages can range from improving company efficiency to boosting employee morale, for example:

- Assists in measuring progress and generates data for future performance comparisons
- Defining specific goals and deadlines to simplify goal management
- Increasing productivity and goal attainment

- Aligning individual goals with the company's overall mission
- Developing a method for companies and workers to hold one another accountable
- Encouraging departmental, team, individual, and stakeholder responsibility and collaboration
- Revealing how company operations impact people and the planet
- Providing a dependable structure for continuous improvement

The implementation of IT sustainability OKRs can be a powerful way for businesses seeking to identify, measure, and achieve sustainability performance objectives. They can also serve as a reliable guide for navigating goals and laying the groundwork for effective team-building and collaboration. When used correctly, sustainability OKRs can maximise the impact of your efforts by focusing each team and employee on the same larger mission.

Investment Decisioning

In most organisations, investment decisions that allocate funds to change activity are focused on growth, cost, operational efficiency, and customer satisfaction and have not been based on sustainability criteria. However, I have started to see the idea of incorporating sustainability into investment decision-making becoming more common in progressive organisations over the past year.

When OKRs for sustainability are used in investment decisions, KPIs are used to track progress, and the IT governance framework is updated to help with execution, this creates a chance to deeply integrate sustainability into all future change activities. Leaders and stakeholders throughout the organisation need to evolve existing decision frameworks in order to meet their Net-Zero commitments, as existing processes and decision criteria are not fit for purpose and can cause delays to a business taking action.

In any large organisation, rapid change will happen almost immediately when a business or technology project no longer receives support, priority, and funding because they do not meet the sustainability criteria of an organisation wide OKR.

4. Operationalise

The People, Process, and Technology (PPT) framework[150] is a well-known method for implementing change in which all three factors are considered and effectively balanced, i.e., employees are able to complete organisational tasks with the assistance of established procedures and access to the appropriate technological resources. I've covered the technology in detail throughout the book thus far, and the remainder of this section will concentrate on the Process elements, leaving the People component as the book's final chapter, as I believe that building a culture of sustainability in IT is so crucial that it merits its own focused discussion.

Process

In the PPT framework, 'Process' describes 'how' something is accomplished. A process is usually defined as a series of stages that can be repeated with identical results expected each time. This also applies when developing and enforcing effective sustainability-oriented procedures. It is important to first concentrate on the processes that we expect to have the greatest impact, and the results from executing these processes need to be measurable.

The following five key processes should be considered when developing a sustainable IT strategy, and each should be supported with a suitable measurement capability in order for it to have the greatest impact.

---

[150] 'Organizational World', Howard J Leavitt, 1973

1. Make decisions based on a sustainable technical architecture

In many organisations, the technical architecture and strategy serve as a blueprint to guide a wide range of important decisions. These can include decisions such as whether to invest in a new idea, select and use new technologies, identify and classify legacy platforms, make choices to replace, remediate, or increase the efficiency of existing components, or whether to modernise, all of which can be decided upon based on their alignment to an agreed-upon technical architecture.

Placing sustainability at the centre of technical architecture and building this into your sustainable IT strategy is a highly effective method of ensuring that all future technology changes are delivering towards your low-carbon goals.

To achieve this, think about the following:

- Requesting that the organisation's architects immediately incorporate sustainability criteria, metrics, and targets into their roadmaps, architecture/design evaluations, and sign-offs.
- Updating or implementing an architecture design review or sign-off process for every major change to alter, upgrade, or remove a component of the architecture to include a consideration of its energy, climate, and environmental impact.

2. Implement DevSustainOps

DevSustainOps is a term I created to connect and leverage the efficiencies, principles, best practices, and automation of DevOps and FinOps in order to accelerate the integration of sustainability into software development and operation. DevSustainOps has been designed to be the intersection between DevOps and FinOps, rather than a 'fifth Op' (the other two being DataOps and SecOps). It can be a powerful way for technology leaders to rapidly introduce sustainability into existing processes that help determine the speed, effectiveness, quality, and cost of IT management.

The basis of DevSustainOps is to assist practitioners by quickly augmenting their DevOps and FinOps activities in order to achieve a low-carbon outcome for the digital products and services they are engineering and managing. The benefits of adopting a DevSustainOps approach are highlighted in the diagram below and summarised as follows:

- To begin a deeper and more complete decarbonisation of the digital products and services being designed, engineered, and operated.
- By using quality management processes, the introduction of sustainability into software delivery becomes measurable and trackable, enabling appropriate prioritisation, spending decisions, and performance reporting.
- DevSustainOps is completely complimentary to and will leverage and add to the operational and cost efficiencies derived from DevOps and FinOps maturity.
- DevSustainOps also provides product management with the ability to deliver lower-carbon digital solutions to customers.

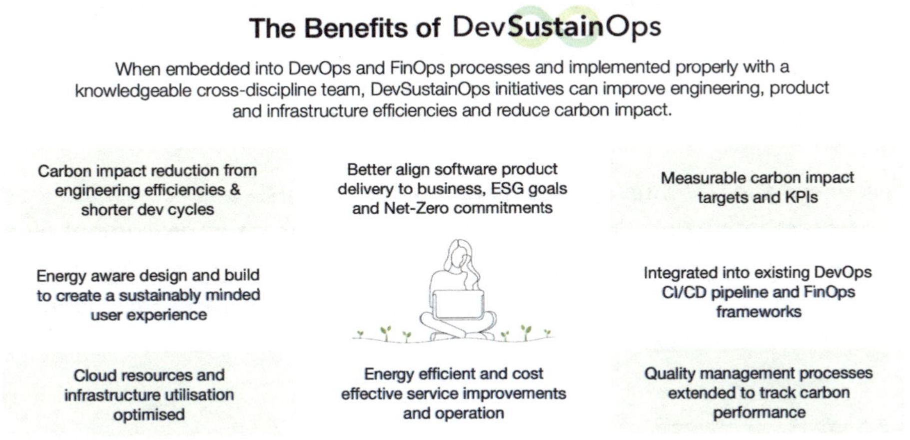

The DevSustainOps principles were developed to be fully compatible with existing DevOps and FinOps frameworks, making their adoption straightforward for any organisation that has already invested in these approaches. The six principles are shown in the diagram below:

The 6 Principles of DevSustainOps

Responsible Collaboration Engineering, operations, finance, business and ESG teams must collaborate and take responsibility for their carbon impact	**Actionable & Agile** Steps identified to reduce energy usage and lower carbon impact are actionable and achievable	**Measured** Measurement based data is the basis for informing actions and decisions to reduce carbon impact
Embedded & Planned Fully integrated into DevOps and FinOps initiatives to constantly maximise impact and value with clear execution plan	**Constant Reduction** Use CI/CD approach and efficiencies combined with FinOps cost management techniques to deliver constant carbon reduction	**Reportable** Routinely produce and share reports and performance against set carbon reduction metrics across the team

The alignment of the DevSustainOps maturity model with DevOps and FinOps maturity assessments allows for rapid adoption and progress monitoring; refer to the maturity framework on the next page. To drive their sustainable IT strategies, technology leaders should integrate this framework into their work practices and develop plans for achieving the Advanced Level. By doing so, DevSustainOps can bring lower-carbon IT and the decarbonisation of digital products and services closer to reality and empower those responsible for delivering new digital products to take the lead in taking action.

Consider taking the following actions to get started:

- Review the DevSustainOps process mapping[151] and conduct a self-assessment against the maturity framework to create improvement plans.
- Obtain a deeper understanding of DevSustainOps by ensuring that your DevOps and FinOps practices are aware of and implement changes to existing processes.

---

151 'DevSustainOps: The management practice of embedding and integrating sustainability into the delivery, development and operation of digital products and services', Version 2.0, March 2023 is free and available using the details at the end of this section.

DevOps	Initial	Managed	Defined	Measured	Optimised
Description	**Not started DevOps** Traditional environment with Dev and Ops separately handled	**Starting DevOps** Beginning of change in mind-set focused on agility in Dev and initial automation in Ops, with emphasis on collaboration	**Fundamental DevOps** Organisation-wide transformation begins with defined processes and automation	**Managed DevOps** Improved understanding of process and automation, the start of continuous improvement	**Optimising DevOps** Visible achievements, strong team collaboration and employee recognition
Characteristics	Manual processes, outages, unplanned work. Tech Debt.	Cultural change started, automated build process. Automated regression suite and alert management.	Cultural change happening, change management process defined. Reuseable scripts, product focused.	Integrated tool chain, SLAs in place. Metrics collected and shared.	Continuous Improvement. Automated incident resolution, virtualisation.

DevSustainOps	Starter	Intermediate	Advanced
Responsible Collaboration	Teams from across technology, business, finance and ESG identified and responsibilities for achieving low carbon targets being set and agreed.	Low carbon targets for technology are intrinsically part of Objective Key Results and KPIs. All teams support each other to ensure investment and prioritisation are aligned.	Continuously improving to reduce carbon impact, increase energy efficiency and support innovation. Third party suppliers and vendors now actively collaborating.
Actionable	Actions are being identified and assessed for impact. Prioritisation of immediately actionable reduction informs initial plan.	Action impacts cover all elements of software engineering, are calculated and being executed.	Pre-build modelling and simulation at design stage prevents inefficiency.
Measured	Carbon Measurement started across all categories of engineering, cloud, hosting and capabilities.	Measurement is granular and detailed, with clear targets set and agreed across the organisation.	Fully integrated and automated measurement, enabling real time decisioning to manage and further reduce carbon costs.
Carbon Reporting	Basic reporting underway.	Reporting is consistent, complete and regularly reviewed.	Detailed Technology decarbonisation reporting forms input to regulated and mandated ESG reporting standards.
Constant Reduction	Culture of constant reduction being defined and syndicated.	A culture of digital decarbonisation forms part of the engineering experience and is valued and rewarded.	Technology decarbonisation leads the organisations efforts on achieving net-zero and creates digital product differentiation for customers.
Embed & Plan	DevOps and FinOps started. Sustainable IT and carbon impact awareness growing but not embedded or fully planned.	Principles of DevSustainOps completely embedded into DevOps and FinOps initiatives. Principles embedded into technology strategy.	Agile delivery, DevOps, FinOps and DevSustainOps are fully embedded into all new product development.

FinOps	Crawl	Walk	Run
Understand Fully Loaded Costs	Visibility into IT spend	Granular cost allocation	Team level budgets and tracking
Cost Drivers & Forecasting	Cost drivers aligned to CSP native services	Understand cost drivers with loosely coupled business goals	Product definitions, business objectives and cost drivers well aligned
Performance Benchmarking	Trending and variance analysis	Internal team benchmarking	Industry peer level benchmarking
Real-time Decision Making	Understand if resources are efficiently provisioned	Remove underutilised services	Automation of resource optimisation
Predict, Plan & Purchase	Rightsizing instances and services	Centralised Reserved Instances (RI) buying process	Compare pricing across CSPs and services
Collaboration	Defined governance & controls for cloud usage	Finance moves at the speed of IT	Continuously improve for efficiency and innovation

3. Introduce sustainable procurement practices

'Sustainable Procurement' is a term that refers to the incorporation of Corporate Social Responsibility (CSR) principles into a company's procurement practices while maintaining compliance with business and stakeholder needs. The following quote provides a good summary of what is becoming expected:

"When I choose a supplier, yes, the supplier has to meet my cost profile, but it also needs to meet my carbon profile, my nature profile and my social profile. And that information flow is going to have to be measured, disclosed and brought through the value chain so these decisions can be informed."

World Business Council for Sustainable Development (WBCSD)

As a core element of any sustainable IT strategy and tackling ICT related scope 3 emissions, all technology leaders should consider their supply chain and purchasing decisions. As we've seen earlier in the book, scope 3 emissions can account for between 70 and 90 percent of a company's total emissions, and the ICT element can be up to 29 percent. As a result, sustainable procurement practices need to form a material and significant part of any action plan.

Most technology leaders will be aware that their IT estate is heavily dependent on third-party IT outsourcing, infrastructure provisioning, software services, hosting, and network capabilities. All of which will be procured and form part of their organisation's scope 3 impact. The energy consumption generated by these procured products and services will have implications across all scope categories; therefore, this has to be a major area of focus and attention if you want to reduce carbon impact.

"Environmental, social, and ethical performance – or Sustainability – is an essential factor for smart business today. More and more companies, including your trading partners, are under immense pressure to improve their transparency and sustainability practices, and this includes the sustainability of their trading partners."

EcoVadis

Going forward, effective sustainable IT procurement practices will need to include carbon-related specifications, targets, and criteria that help achieve environmental protection. A commitment to consistently executing sustainable procurement choices will ensure that a company's core sustainability values can be reflected throughout the entire life cycle of its products and services.

To implement sustainable procurement practices in your organisation, some immediate next steps might be to:

- Adopt practices to future-proof your company's sustainable procurement and investments by implementing policies that increase the long-term viability of all procured products and services.
- Implement a consistent scoring mechanism to evaluate and compare your value chain objectively, and promote the use of standard measurement tools to evaluate the carbon and environmental impact of your internal IT estate and external service providers. For example, I have written a 'Low Carbon Software Specification' that can help any organisation assess the carbon emission impact of the software products or services they are selecting to purchase. It is easier to make smart choices on new products before purchasing than to remediate the carbon impact once you have installed, configured, and are using them.
- Utilise the IT purchasing power of your organisation to ensure that all vendors are evaluated consistently and systematically as part of all future Requests for Information and/or Pricing (RFI/RFP). Inserting the Low Carbon Software Specification at this stage is highly effective and will encourage your software vendors to demonstrate their product decarbonisation activities. I'll give you access to the specification at the end of this chapter.

4. Make sustainability a Non-Functional Requirement

As previously explained in the Design chapter, Non-Functional Requirements (NFRs) are a powerful process for helping define system attributes such as security, reliability, performance, maintainability, scalability, and usability. They function as persistent constraints or restrictions on system design across multiple backlogs, and NFRs are as essential to agile delivery as functional epics, capabilities, features, and stories. Sustainability should be regarded as a crucial NFR, with implications across all other requirements.

Similar to DevSustainOps and the Low Carbon Software Specification, I have written a Low Carbon Non-Functional Requirement that helps identify the parameters needed to cover the production and use lifecycle stages for any software product being designed, built, and operated. This then sets the metrics and quality attributes needed to achieve the NFR on a measurable basis. At the end of this section, I will provide details on how to access all of these resources for free as accelerators to get you started on your decarbonisation journey.

5. Set up a Sustainable IT Centre of Excellence

A centre of excellence (CoE) is established when a group of experienced knowledge workers come together to set up best practices in a specific area of interest. The idea of forming special interest groups for thought leadership originated in lean manufacturing and is now widely adopted across many sectors.

CoEs are typically established when there is a knowledge or talent gap within an organisation. They are perfectly structured to facilitate the rapid implementation of all the sustainability-based process enhancements outlined above. COEs can help eliminate inefficiency and advance an organisation to the next level of a maturity model; for instance, they would be a highly effective method for accelerating DevSustainOps adoption and execution.

A Sustainable IT CoE may be permanent or temporary in nature. When a CoE is permanent, team members typically have additional responsibilities; however, when a CoE is temporary, team members may be relieved of their normal responsibilities for the duration of the CoE's operation. CoE members should, within the context of a sustainable IT strategy, be responsible for disseminating knowledge about sustainable IT, managing data, and assisting teams in implementing recommended best practices. If you are thinking about setting up a CoE, consider the following:

- Once established, empower the CoE with the authority to push for the adoption of consistent practices, processes, measurement tools, and data in support of your sustainable IT strategy.
- Have the CoE set and measure objectives and metrics.
- The CoE should assist in accelerating data standards and reporting in order to inform decisions and monitor their progress.
- It should demonstrate to the organisation that sustainable IT is integral to the overall performance and objectives of the technology organisation.

Supporting your Sustainable IT Strategy

Once all four best practices are in place, the foundation is set for establishing a successful sustainable IT strategy that will contribute towards achieving your organisation's Net-Zero commitments. It will be a strategy that will be governed, executed, and operationalised in accordance with clearly defined and measurable performance targets.

Alongside all of the GoCodeGreen resources you have used throughout this book, you can access my DevSustainOps framework, Low Carbon Software Specification, and Low Carbon Non-Functional Requirements by visiting the resources section at www.decarbonise.digital or using the QR code above.

New Reporting Standards and the Impact on IT

"There is an ever-growing list of rules and regulations for enterprises to get their heads around when it comes to sustainability. Enterprises are under pressure, or even mandated, to incorporate Corporate Sustainability Reporting Directives (CSRD) into their regulatory landscape."

PA Consulting & GoCodeGreen[152]

If the reasons for decarbonising digital aren't already clear, upcoming reporting standards and regulations may serve as the final impetus for taking action. The Corporate Sustainability Reporting Directive (CSRD) requires companies to report on the impact of their activities on the environment and society, as well as to conduct an audit of the information reported. This legislation is expected to be fully implemented in 2024.

Companies will be required to provide substantially more detail when reporting their environmental impact, and those that fail to do so will be subject to audits and fines.

The European Union (EU) adopted this initiative because it is in line with its commitments under the European Green Deal. This acknowledges that climate change and environmental degradation have a negative impact on our planet and seeks to raise the bar for businesses. The EU has pledged to become a more resource-efficient and environmentally conscious economy, with zero net emissions of greenhouse gases by 2050 and a 50 percent reduction of these emissions by 2030.

---

152 'IT Sustainability Think Tank: Keeping on top of the regulatory landscape', Richard Barnes-Webb (PA Consulting) and Eric Zie (GoCodeGreen), Computer Weekly, July 2023

The CSRD acknowledges that businesses contribute significantly to environmental degradation and waste, and the EU believes that the largest contributors must be held accountable. In addition to compelling companies to reduce their emissions, the CSRD will assist investors, consumers, and other stakeholders in evaluating non-financial aspects of company management. This will enable consumers and investors to advocate through their purchasing choices for green practices and encourage companies to prioritise a sustainable approach to their business practices.

In addition to and in support of the CSRD, the use of common standards will ensure that companies across the EU report comparable and trustworthy information on sustainability. Companies that are required by the Accounting Directive to report sustainability information must use both the International Financial Reporting Standards (IFRS) and the European Sustainability Reporting Standards (ESRS) across the EU. By avoiding the use of multiple voluntary standards and fostering a culture of greater public accountability, these common standards are anticipated to help companies reduce their reporting costs over the medium and long term.

The International Sustainability Standards Board (ISSB) and the Global Reporting Initiative (GRI) have established international standards, which the ESRS complies with. Companies required to report in accordance with the ESRS on climate change will, for the most part, disclose the same information as those using the ISSB standard for climate-related disclosures. ESRS-mandated climate change disclosures will provide additional information on impacts relevant to users other than investors, such as business partners, trade unions, social partners, and academics.

With the adoption of the ESRS, the EU goes further than any other major jurisdiction in incorporating ISSB standards into its own legal framework. By doing so, the EU is contributing to the development of a coherent global framework and the global comparability of reported sustainability information. Under the ESRS, five environmental standards exist. Each standard adheres to the same underlying premise: that each company must disclose all material risks, impacts, and opportunities. Then, companies should disclose the policies, actions, and objectives they have in place to mitigate these risks and impacts. The five standards can be summarised as follows:

- ESRS E1: The Climate Change Standard consists of scope 1, 2, and 3 emissions, climate-related hazards, carbon pricing, energy mix, and a credible transition plan to align with the Paris Agreement.
- ESRS E2: The Pollution Standard requires disclosure of air, soil, and water pollution caused by direct operations and operations in the value chain.
- ESRS E3: The Water and Marine Resources Standard addresses water consumption and recycled or reused water, in addition to any significant negative effects on marine ecosystems.
- ESRS E4: The Biodiversity and Ecosystems Standard, requires companies to disclose their impacts on the natural environment as well as a transition plan to address biodiversity loss.
- ESRS E5: The Resource Use and Circular Economy Standard, requires companies to disclose relevant circular material resource flows and waste categories generated by their operations.

Emissions disclosures serve as the basis for the majority of climate reporting standards and have become 'table stakes' for corporate climate programmes. It's for good reason: even if a company has a light physical supply chain, for example, if it's a software company, it may still emit significant greenhouse gases through activities such as cloud usage and the purchase of other products and services.

The majority of businesses will be required to report on E1, the Climate Change Standard, which includes greenhouse gas emissions. In addition to their climate hazards assessment and any policies related to climate change mitigation and adaptation, organisations will be required to disclose their full scope 1, 2 and 3 emissions.

To summarise:

- CSRD is mandatory.
- CSRD introduces 12 standards.
- CSRD focuses on double materiality, recognising the importance of both financial and non-financial aspects.

- GRI is voluntary.
- GRI provides a wide range of reporting options and disclosures.
- GRI focuses on impact materiality.

- ISSB can be mandated and combined with jurisdiction-specific requirements.
- ISSB focuses on general reporting requirements and climate-related disclosures.
- ISSB focuses on financial materiality.

It may not be immediately apparent, but the implications of the last few pages are huge. Over the remainder of this section, I will explain what this means for digital and why all IT leaders need to start preparing NOW.

Technology leaders: start preparing NOW

What do these reporting requirements mean for technology leaders? As regulations are expected to become more elaborate and detailed, with increasing requirements for reporting, tracking, and monitoring of sustainability metrics, technology leaders will undoubtedly experience increased scrutiny within their organisations as well as externally.

Non-compliance with CSRD will result in sanctions, but the actual cost of non-compliance will be much higher. Customers will evaluate prospective suppliers through the lens of CSRD, and they will choose to partner with organisations that have a higher sustainability rating. On top of being more aware of the impact their purchases have on the environment, consumers will avoid brands that neglect sustainability initiatives. Non-compliance with CSRD could destroy the image and credibility of a brand. Ultimately, if businesses do not take sustainability seriously, they run the risk of losing employees, consumers, and public trust.

Reporting standards: driving action in IT

The new reporting standards shouldn't be viewed negatively. They should be seen as catalysts for positive change, despite the work that will be required to comply with them. After an evaluation of these new standards, I have identified a range of potential repercussions for IT leaders, and I'll cover these in more detail later in this chapter. I'll be focusing on the following three standards where I can see clear IT-related implications:

IFRS S1 (International Financial Reporting Standards): General Requirements

That requires an entity to disclose information about its sustainability-related risks and opportunities that is useful to users of general-purpose financial reports in making resource allocation decisions.

IFRS S2: Climate-related Disclosures

That requires an entity to disclose information about its climate-related risks and opportunities that is useful to consumers of financial reports for general purposes in making decisions about providing resources to the entity.

ESRS E1 (in draft)

The Climate Change Standard is one of five environment-specific standards of the European Sustainability Reporting Standards (ESRS), developed by the European Financial Reporting Advisory Group (EFRAG) and to be deployed in accordance with the CSRD.

Key reporting themes

These are new standards, so for a while, it may remain challenging to understand how they will be fully implemented. Their introduction will require businesses to combine sustainability-related disclosures and financial statements in the same reporting package, so we already know that they will be on a repeated, annual cycle. Adapting to them will require collecting new climate-related data and compiling it in accordance with the new standards' deadlines, reporting processes, and compliance requirements. Despite the significance of these standards, there is currently no clear or formal guidance on how IT teams should interpret and implement them. What follows is my 'unofficial' perspective on this challenge, built around four recurring themes that emerged as I reviewed them:

1. Materiality

In the context of IFRS[153], materiality is defined as "information that is material if omitting, misstating, or obscuring information that could reasonably be expected to influence decisions and which provides information about a specific reporting entity." For many businesses undergoing digital transformation or dependent on digital to deliver their products and services to customers, it is easy to argue that ICT is material to financial performance. If we also recognise that the carbon and environmental impacts of using IT are substantial, then compliance with the new reporting standards will be required.

2. Strategy, risks, and opportunities

Per the IFRS S1 standard, "The objective of sustainability-related financial disclosures on strategy is to enable users of general-purpose financial

---

[153] per IFRS S1 'General Requirements for Disclosure of Sustainability-related Financial Information', June 2023

reports to understand an entity's strategy for managing sustainability-related risks and opportunities." The significance of ICT on an organisation's financial performance and environmental impact means that technology-dependent businesses will need to assess sustainability-related risks and opportunities associated with their use of ICT and reliance on digital.

3. Metrics and Targets

Metrics and targets repeatedly appear as a theme across all three standards. According to IFRS S2, the way in which measurement and monitoring of performance for sustainability-related risks and opportunities is undertaken is vitally important: "to understand an entity's performance in relation to its climate-related risks and opportunities, including progress towards any climate-related targets it has set and any targets it is required to meet by law or regulation". The standards contain specific disclosures that require an IT response based on the materiality, risks, and opportunities associated with the use and impact of technology.

4. Scope 3, business model, value chain, and technology

There are references to scope 3, business model, and value chain across each of the standards, as well as specific references to technology within ESRS E1. For each reporting standard, there is now a strong connection to ICT, per the following examples:

Scope 3

"The scope 3 category is one of the 15 types of scope 3 emissions identified by the GHG Protocol Corporate Standard and detailed by the GHG Protocol Corporate Value Chain (Scope 3) Accounting and Reporting Standard" (source: ESRS E1).

As we have seen, up to 29 percent of an organisation's scope 3 impact can be ICT-related, so the reporting requirement to understand the contribution of these emissions across the relevant scope 3 categories will require IT leaders to identify and measure the source (through their supply chain) and scale of these impacts, along with associated actions to remove them in line with their organisation's Net-Zero commitments.

Business model

"An entity's system of transforming inputs through its activities into outputs and outcomes that aims to fulfil the entity's strategic purposes, create value for the entity, and hence generate cash flows over the short, medium, and long term" (source: IFRS S1/S2).

For any organisation that is digital first or digitally dependent on creating and distributing its products and services to customers, this reporting standard will apply to its use of ICT. In these contexts, technology will be key to fulfilling the organisation's strategic purpose, creating value, and generating revenue.

Value chain

"The full range of interactions, resources, and relationships related to a reporting entity's business model and the external environment in which it operates. A value chain encompasses the interactions, resources, and relationships an entity uses and depends on to create its products or services from conception to delivery, consumption, and end-of-life, including interactions, resources, and relationships in the entity's operations, such as human resources; those along its supply, marketing, and distribution channels, such as materials and service sourcing and product and service sale and delivery; and the financing, geographical, geopolitical, and regulatory environments in which the entity operates" (source: IFRS S2).

For most organisations, the majority of their ICT is provided through their value chain. This includes the obvious things, like the desktop computers or laptops being used, through to IT services covering resources undertaking software development and supporting operations, to SaaS solutions and the cloud. The previously quoted 29 percent of scope 3 emissions makes ICT a material source of emissions that will need to be understood and reported against at the activity levels stated in the reporting standard above.

Disclaimer for the next section

Before moving on to the detailed reporting disclosures in the following table, I want to reiterate that what follows is my personal interpretation of the new reporting standards and is not intended as guidance. The purpose is to demonstrate how the new and upcoming reporting standards will begin to demand consideration of ICT and digital as material contributors to the overall sustainability of an organisation.

Reviewing the standards

Extract from Sustainability Reporting Standards (source as indicated)	IT/Technology Observations and Considerations
Materiality per ISSB IFRS S1 *Point 17, page 8* *"An entity shall disclose material information about the sustainability-related risks and opportunities that could reasonably be expected to affect the entity's prospects."*	Any company that relies heavily on IT and technology to create and deliver products and services to customers should evaluate whether the climate and environmental impact of its operations are material to its overall sustainability goals.
Sustainability-related risks and opportunities per IFRS S1 *Point 30, page 11* *"An entity shall disclose information that enables users of general purpose financial reports to understand the sustainability-related risks and opportunities that could reasonably be expected to affect the entity's prospects. Specifically, the entity shall:*	If a company's IT-related activities have a significant impact on its future prospects, it must measure the climate and environmental impact of its use of technology to support its business model and through its value chain, as well as the associated scope 3 emissions.

(a) describe sustainability-related risks and opportunities that could reasonably be expected to affect the entity's prospects;

(b) specify the time horizons–short, medium or long term–over which the effects of each of those sustainability-related risks and opportunities could reasonably be expected to occur; and

(c) explain how the entity defines 'short term', 'medium term' and 'long term' and how these definitions are linked to the planning horizons used by the entity for strategic decision-making."

Where these emissions are material, the business will need to consider how the associated sustainability risks and opportunities are measured and managed and how they contribute to the business's overall risk management time horizons.

This will require the IT function to establish a baseline, identify the sources of sustainability risk and opportunity, and devise a plan to manage them.

Business model and value chain per IFRS S1

Point 32, page 11
An entity shall disclose information that enables users of general purpose financial reports to understand the current and anticipated effects of sustainability-related risks and opportunities on the entity's business model and value chain. Specifically, the entity shall disclose:

(a) a description of the current and anticipated effects of sustainability-related risks and opportunities on the entity's business model and value chain; and

(b) a description of where in the entity's business model and value chain sustainability-related risks and opportunities are concentrated (for example, geographical areas, facilities and types of assets).

Where IT has a material impact on the business's prospects or presents substantial sustainability risks and opportunities, the IT function will be required to compile data for this disclosure.

Being able to measure and produce data will be needed where a large percentage of IT is associated with the value chain (for example, related to third-party managed services, software products, SaaS solutions, or the use of third-party cloud or hosting services) and dispersed across different geographic locations.

Metrics and targets per IFRS S1

Point 45, page 15
"The objective of sustainability-related financial disclosures on metrics and targets is to enable users of general purpose financial reports to understand an entity's performance in relation to its sustainability-related risks and opportunities, including progress towards any targets the entity has set, and any targets it is required to meet by law or regulation."

Point 51, page 16
"An entity shall disclose information about the targets it has set to monitor progress towards achieving its strategic goals, and any targets it is required to meet by law or regulation."

Point 52, page 16
"The definition and calculation of metrics, including metrics used to set the entity's targets and monitor progress towards reaching them, shall be consistent over time."

Where IT and technology are material to the business entity's prospects, then establishing appropriate metrics and targets to measure performance against sustainability-related risks and opportunities will be required.

An IT function will need to consider what to measure, establish targets that are aligned with the goals of the organisation as a whole, and have a consistent approach to monitoring and continuing measurement over time.

This may include metrics based on energy consumption, CO_2e reduction, carbon intensity, or quantitative measures derived from ratings. Verification of the data used for measurement and subsequent metrics will be required, so IT functions should consider how best to accomplish this as opposed to investing in and employing their own methods.

Sustainability-related risks and opportunities per IFRS S1

Point B4, page 26
"Resources and relationships that an entity depends on and affects by its activities and outputs can take various forms, such as natural, manufactured, intellectual, human, social or financial. They can be internal–such as the entity's workforce, its know-how or its organisational processes–or they can be external –such as materials and services the entity needs to access or the relationships it has with suppliers, distributors and customers."

IT functions must evaluate the following two factors:

Identifying the key external relationships across its value chain that drive a sustainability-related risk or opportunity will require measurement and it should form part of an overall impact assessment.

Point B5, page 26 *"An entity's dependencies and impacts are not limited to resources the entity engages with directly, and to the entity's direct relationships. Those dependencies and impacts also relate to resources and relationships throughout the entity's value chain."*	It is important that activities and outputs have an intellectual and human dimension. This will align with the people and capabilities of the IT function and the evaluation of their ability to mitigate sustainability risks or seize opportunities.
Verifiability per IFRS S1 *Point D21, page 40* *"Verifiability helps to give users confidence that information is complete, neutral and accurate. Information is verifiable if it is possible to corroborate either the information itself or the inputs used to derive it. Verifiable information is more useful to primary users than information that is not verifiable."*	When determining a response and contribution to disclosures, the IT function must consider the 'verifiability' of measurement, metrics, and performance against targets. This may necessitate the participation of both internal (via the ESG function or an aligned Centre of Excellence) and external (auditor or external ESG or comparable specialist organisation) parties.
Strategy per IFRS S2 *Point 8, page 7* *"The objective of climate-related financial disclosures on strategy is to enable users of general purpose financial reports to understand an entity's strategy for managing climate-related risks and opportunities."* ***Climate-related risks and opportunities*** *Point 10, page 7* *"An entity shall disclose information that enables users of general purpose financial reports to understand the climate-related risks and opportunities that could reasonably be expected to affect the entity's prospects. Specifically, the entity shall:*	Similar to the financial and sustainability-oriented disclosure requirements of IFRS S1, the S2 standard focuses on risks and opportunities related to climate change. IT functions will need to consider how the data, measurement, metrics, and targets relate specifically to climate impact; similarly, if the use of IT is material to the business, then understanding the carbon and environmental impact and mapping to climate-related risks and opportunities will be required.

(a) describe climate-related risks and opportunities that could reasonably be expected to affect the entity's prospects"	The risks and opportunities must be focused on both the transition to low-carbon IT solutions and the tangible impacts of climate change caused by IT use.
Business model and value chain *Point 13, page 8* *"An entity shall disclose information that enables users of general purpose financial reports to understand the current and anticipated effects of climate- related risks and opportunities on the entity's business model and value chain. Specifically, the entity shall disclose:* *(a) a description of the current and anticipated effects of climate-related risks and opportunities on the entity's business model and value chain; and* *(b) a description of where in the entity's business model and value chain climate-related risks and opportunities are concentrated (for example, geographical areas, facilities and types of assets)."*	Again, in accordance with the business model and value chain comments for IFRS S1, the IT function will need to identify and consistently measure performance targets for climate-related risks and opportunities. The IT function will need to determine how best to acquire the data required for these evaluations and should use its procurement processes to ensure that its supply chain vendors provide the necessary data to enable qualitative responses.
Strategy and decision-making *Point 14, pages 8-9* *"An entity shall disclose information that enables users of general purpose financial reports to understand the effects of climate-related risks and opportunities on its strategy and decision-making. Specifically, the entity shall disclose:* *(a) information about how the entity has responded to, and plans to respond to, climate-related risks and opportunities in its strategy and decision-making, including how the entity plans to achieve any climate-related targets it has set and any targets it is required to meet by law or regulation. Specifically, the entity shall disclose information about:*	This disclosure implies that the organisation must be able to quantify the climate risks associated with its business and that the IT function must understand how it will contribute to the organisation's plans to achieve climate-related targets. As a result, the IT function must also be able to measure the impact of the actions it can take to resolve climate-related risks and opportunities for each of the disclosure items (i) to (v). For instance, consideration must be given to:

(i) current and anticipated changes to the entity's business model, including its resource allocation, to address climate-related risks and opportunities (for example, these changes could include plans to manage or decommission carbon-, energy- or water-intensive operations; resource allocations resulting from demand or supply-chain changes; resource allocations arising from business development through capital expenditure or additional expenditure on research and development; and acquisitions or divestments);

(ii) current and anticipated direct mitigation and adaptation efforts (for example, through changes in production processes or equipment, relocation of facilities, workforce adjustments, and changes in product specifications);

(iii) current and anticipated indirect mitigation and adaptation efforts (for example, through working with customers and supply chains);

(iv) any climate-related transition plan the entity has, including information about key assumptions used in developing its transition plan, and dependencies on which the entity's transition plan relies; and

(v) how the entity plans to achieve any climate-related targets, including any greenhouse gas emissions targets."

- Impact of retiring energy-hungry applications and IT infrastructure
- The effectiveness of hosting and processing services' energy and water consumption for cooling
- Effectiveness of the IT supply chain
- Possibilities for mitigating climate impacts via workforce and location strategies
- Impact of enhancing digital product and service productivity via specification, design, and build
- A transparent plan that establishes baselines and can measure performance towards meeting climate-related goals, including GHG emission targets

This list starts to feel like a call to action to develop a sustainable IT strategy and a measurable execution plan.

Metrics and targets per IFRS S2 *Point 27, page 14* *"The objective of climate-related financial disclosures on metrics and targets is to enable users of general purpose financial reports to understand an entity's performance in relation to its climate-related risks and opportunities, including progress towards any climate-related targets it has set, and any targets it is required to meet by law or regulation."* *Point 28, page 8* *"To achieve this objective, an entity shall disclose:* *(b) industry-based metrics that are associated with particular business models, activities or other common features that characterise participation in an industry."* ***Climate-related targets*** *Point 33, page 16* *"An entity shall disclose the quantitative and qualitative climate-related targets it has set to monitor progress towards achieving its strategic goals, and any targets it is required to meet by law or regulation, including any greenhouse gas emissions targets."*	Building on the considerations outlined in the Metrics and Targets section of IFRS S1, this 'industry-based' requirement may evolve and attain greater levels of detail and sophistication over time for technology and software companies in particular. Setting metrics and targets associated with 'features that characterise participation in an industry' may also result in more granular disclosure requirements for digital-first businesses. Common characteristics could include, for instance, the environmental impact of data centre and cloud usage, distribution via networks, or even the performance of the development of digitally based products and services.
Scope 3 measurement framework per IFRS S2 *Point B38, page 33* *"An entity's measurement of Scope 3 greenhouse gas emissions is likely to include the use of estimation rather than solely comprising direct measurement. In measuring Scope 3 greenhouse gas emissions an entity shall use a measurement approach, inputs and assumptions that result in a faithful representation of this measurement."*	Based on the challenges involved in direct measurement at the source of emission, it should be expected that scope 3 measurement methods be flexible. This is especially true for ICT, given the complexity of the value chain. Finding a standardised method of calculation will be essential for all IT operations over time.

Point B48, page 35
"Primary data for Scope 3 greenhouse gas emissions includes data provided by suppliers or other entities in the value chain related to specific activities in an entity's value chain. For example, primary data could be sourced from meter readings, utility bills or other methods that represent specific activities in the entity's value chain. Primary data could be collected internally (for example, through the entity's own records), or externally from suppliers and other value chain partners (for example, supplier-specific emission factors for purchased goods or services). Data from specific activities within an entity's value chain provides a more accurate representation of the entity's specific value chain activities and, therefore, will provide a better basis for measuring the entity's Scope 3 greenhouse gas emissions."

Point B49, page 35
"Secondary data for Scope 3 greenhouse gas emissions is data that is not obtained directly from specific activities within an entity's value chain. Secondary data is often supplied by third-party data providers and includes industry-average data (for example, from published databases, government statistics, literature studies and industry associations). Secondary data includes data used to approximate the activity or emission factors. Additionally, secondary data includes primary data from a specific activity (proxy data) used to estimate greenhouse gas emissions for another activity. If an entity uses secondary data to measure its Scope 3 greenhouse gas emissions, it shall consider the extent to which the data faithfully represents the entity's activities."
"Timely data that faithfully represents the jurisdiction of, and the technology used for, the value chain activity and its greenhouse gas emissions"

IT functions must take into account both the need for primary and secondary data to calculate scope 3 emissions and the impact of their value chain. IT teams should seek out primary data sources whenever possible to improve precision, and they should consider how to evaluate their suppliers and the veracity of the data they provide.

Consideration must also be given to the specificity and consistency of activity metrics in the value chain, and IT teams should select tools, methods, and approaches to support the capture of this data.

Verified data *Point B53, page 36* *"An entity shall prioritise Scope 3 greenhouse gas emissions data that is verified. Verification can provide users of general purpose financial reports with confidence that the information is complete, neutral and accurate."*	Similar to IFRS S1 comments on 'verifiability,' a particularly important consideration for IT functions is who and how to verify IT-related data most effectively. It is possible that a combination of internal, external, and auditing services will be necessary to meet the standard.
Disclosure Requirement E1-1 – Transition plan for climate change mitigation per ESRS E1 *Point 15, page 6* *"(b) by reference to GHG emission reduction targets (as required by Disclosure Requirement E1-4) and the climate change mitigation actions (as required by Disclosure Requirement E1-3), an explanation of the decarbonisation levers identified, and key actions planned, including changes in the undertaking's product and service portfolio and its adoption of new technologies"*	The idea of decarbonisation levers (defined below) in relation to climate mitigation and GHG emission reduction is extremely important. The decarbonisation of digital and associated actions outlined in this book should be regarded as one of the most important levers for digital-first and ICT-reliant organisations.
Disclosure Requirement E1-3 – Actions and resources in relation to climate change policies per ESRS E1 *Point 24, page 8* *"In addition to [draft] ESRS 2 DC-A, the undertaking shall:* *(a) when listing key actions taken in the reporting year and planned for the future, present the climate change mitigation actions by decarbonisation lever including the nature-based solutions;* *(b) when describing the outcome of the actions for climate change mitigation, include the achieved and expected GHG emission reductions; and* *(c) relate significant monetary amounts of CapEx and OpEx required to implement the actions"*	Digital decarbonisation and sustainable IT as a 'decarbonisation lever': In this context, the standards will require the actions to be defined, the outcomes and GHG emission reductions to be measured and monitored, and the implementation costs to be stated. This will require the IT leader to formalise and structure a clearly defined sustainable IT strategy.

Metrics and targets
Disclosure Requirement E1-4 – Targets related to climate change mitigation and adaptation per ESRS E1

Point 32, page 9
"If the undertaking has set GHG emission reduction targets, [draft] ESRS 2 DC-T and the following requirements shall apply:

(a) GHG emission reduction targets shall be disclosed in absolute value (either in tonnes of CO2 equivalent or as a percentage of the emissions of a base year) and, if deemed meaningful, in intensity value;
(b) GHG emission reduction targets shall be disclosed for Scope 1, 2, and 3 GHG emissions. The undertaking shall explain how the consistency of these targets with the GHG inventory boundaries is ensured (as required by Disclosure Requirement E1-6). The undertaking shall not include GHG removals, carbon credits or avoided emissions as a means of achieving the GHG emission reduction targets;
(c) the undertaking shall disclose its current base year and baseline value, and from 2030 onwards, update the base year for its GHG emission reduction targets after every five-year period thereafter. The undertaking may disclose the past progress made in meeting its targets before its current base year provided that this information is consistent with the requirements of this [draft] Standard;
(d) GHG emission reduction targets shall at least include target values for the year 2030 and, if available, for the year 2050. From 2030, target values shall be set after every five-year period thereafter;"

Accepting that digital decarbonisation is a 'decarbonisation lever', this standard clarifies the importance of measuring the GHG emission reduction targets associated with an IT function's sustainable IT strategy. Disclosure will be based on tCO2e and carbon intensity reductions, covering scopes 1, 2, and 3, and will need to be baselined and regularly measured over time in order to meet predetermined carbon emission goals.

Disclosure Requirement E1-6 – Gross Scopes 1, 2, 3 and Total GHG emissions per ESRS E1 *Point 42, page 11* *"(c) gross Scope 3 GHG emissions as required by paragraph 41(c) is to provide an understanding of the GHG emissions that occur in the undertaking's value chain beyond its Scope 1 and 2 GHG emissions. For many undertakings, Scope 3 GHG emissions may be the main component of the GHG inventory and are an important driver of the undertaking's transition risks.* *(d) total GHG emissions as required by paragraph 41(d) is to provide an overall understanding of the undertaking's GHG emissions and whether they occur from its own operations or the value chain. This disclosure is a prerequisite for measuring progress towards reducing GHG emissions in accordance with the undertaking's climate-related targets and EU policy goals."*	This standard disclosure reinforces the need for scope 1, 2, and 3 emission measurement for IT functions and emphasises the importance of scope 3 emissions in the value chain.
Appendix A: Defined Terms Decarbonisation levers per ESRS E1 *Page 17* *"Aggregated types of mitigation actions such as energy efficiency, electrification, fuel switching, use of renewable energy, products change, and supply-chain decarbonisation that fit with undertakings' specific actions."*	With regard to technology, a decarbonisation lever centred on sustainable IT would include a variety of actions promoting energy efficiency, the use of renewables, product selection, and supply chain decarbonisation across the digital products and services being developed and operated by the business.

Disclosure Requirement related to [draft] ESRS 2 IRO-1 Description of the processes to identify and assess material climate-related impacts, risks and opportunities per ESRS E1

AR10, page 23
"When disclosing the information on the processes to identify and assess climate impacts as required under paragraph 18(a), the undertaking shall explain how it has:
(a) screened its activities and plans in order to identify actual and potential future GHG emission sources and, if applicable, drivers for other climate-related impacts"

AR13, page 23
"When disclosing the information on the processes to identify transition risks and opportunities as required under paragraph 18(c), the undertaking shall explain whether and how it has:
(a) identified transition events (see the stated technology examples below) over the short-, medium- and long-term time horizons and screened whether its assets and business activities may be exposed to these events."

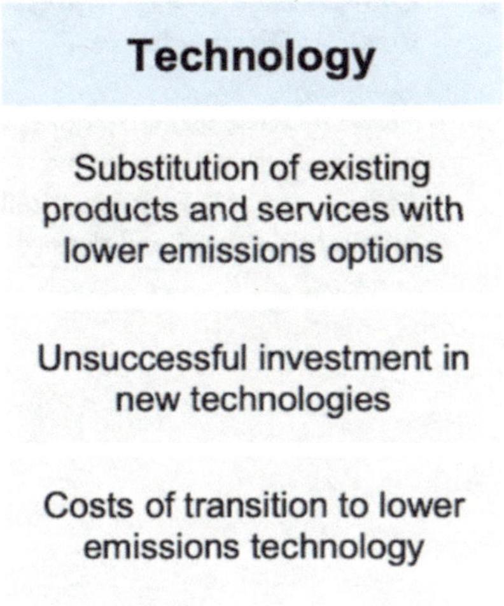

AR10: IT functions must consider how they screen activities across the use of ICT. Depending on the breadth and complexity of the business entity's IT environment, various levels of software and hardware screening may be required.

First, the standard's specific mention of technology is revealing. This can range from carbon capture technologies to ICT technologies, such as cloud computing or more effective software applications.

IT departments must consider the technology landscape, innovation, and broader FinTech ecosystem in order to map time horizons and identify relevant business transition events. These should be monitored as part of the sustainable IT strategy.

Disclosure Requirements E1-3 – Actions and resources in relation to climate change policies per ESRS E1

AR20, page 25
"(b) aggregate types of mitigation actions (decarbonisation levers) such as energy efficiency, electrification, fuel switching, use of renewable energy, products change, and supply-chain decarbonisation that fit the undertakings' specific actions"

AR20: requires IT functions to be able to aggregate and deduplicate their decarbonisation outcomes across all of their actions. To effectively support the sustainable IT decarbonisation lever, a formal structure, reporting, and governance approach will be required.

Disclosure Requirements E1-6 – Gross Scopes 1, 2, 3 and Total GHG emissions per ESRS E1

AR44, page 33
"(c) screen its total Scope 3 GHG emissions based on the 15 Scope 3 categories identified by the GHG Protocol Corporate Standard and GHG Protocol Corporate Value Chain (Scope 3) Accounting and Reporting Standard (Version 2011) using appropriate estimates;

(f) at least update the Scope 3 inventory every three years and in case of major changes (Explanatory note: Scope 3 GHG emissions need to be disclosed every year, but a full update of the underlying inventory is only expected every three years unless major changes occur)"

AR44: requires IT functions to be able to screen and differentiate across the 15 scope 3 categories relevant to ICT. This will be challenging in the short term as the value chain will need to improve its ability to map data and adhere to GHG Protocol standards. IT teams ought to maintain awareness of the disclosure requirement and exert pressure on their supply chains to assist with compliance.

AR49, page 36
"If it is material for the undertaking's Scope 3 emissions, it shall disclose the GHG emissions from purchased cloud computing and data centre services as a subset of the overarching Scope 3 category "upstream purchased goods and services".

AR49: this disclosure demonstrates an awareness of the growing significance of ICT by specifically highlighting cloud and data centre services.

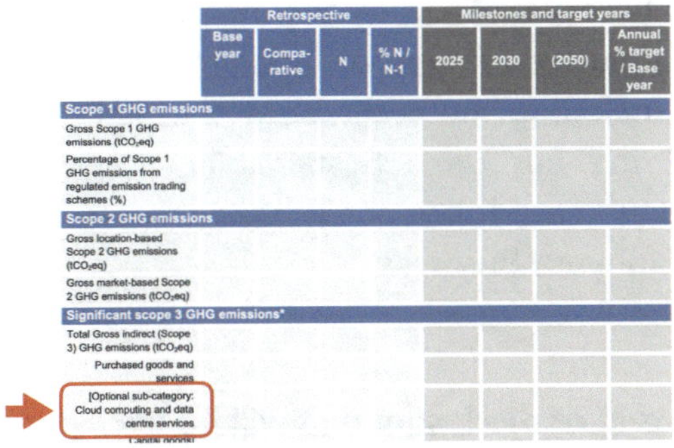

	Retrospective				Milestones and target years			
	Base year	Compa-rative	N	% N / N-1	2025	2030	(2050)	Annual % target / Base year
Scope 1 GHG emissions								
Gross Scope 1 GHG emissions (tCO_2eq)								
Percentage of Scope 1 GHG emissions from regulated emission trading schemes (%)								
Scope 2 GHG emissions								
Gross location-based Scope 2 GHG emissions (tCO_2eq)								
Gross market-based Scope 2 GHG emissions (tCO_2eq)								
Significant scope 3 GHG emissions*								
Total Gross indirect (Scope 3) GHG emissions (tCO_2eq)								
Purchased goods and services								
[Optional sub-category: Cloud computing and data centre services								

The key for IT functions is to look beyond simple compute-related impact to understand what originally drove the demand for that compute. The answer is software. Increasing the energy efficiency of software decreases the demand for hardware and other elements of the technology value chain; therefore, this standard requires a more holistic, all-encompassing, approach to measurement and action.

Ask yourself 3 questions...

Question 1

Are you a digital-first business? Or invested in a digital transformation? Does your business depend on technology? Are many of the products and services you offer to clients digitally-based? And are they delivered to customers through digital channels?

If you answered yes, it is likely that your reliance on ICT will be considered material under the new reporting standards[154] because it has a direct impact on both the financial prospects of your business and technology's role as a key enabler in meeting your organisation's sustainability commitments and goals.

Question 2

Are you aware that technology has a significant impact on carbon emissions and the environment? And did you know that a significant portion of your technological use contributes to your scope 3 emissions?

---

154 International Financial Reporting Standard, S1 - General Requirements for Disclosure of Sustainability-related Financial Information

If yes, and based on the material nature of technology in your business, then it is likely that under new reporting standards[155], you will need to disclose the climate-related impact of your IT, including the technology that is in your value chain. Under the terms of the European Sustainability Reporting Standard[156], 'digital decarbonisation' could be considered one of the significant 'Decarbonisation Levers' for your business.

Question 3

Given this, are you aware of the extent of disclosure that will be necessary? Do you feel prepared and able to respond?

If not, you are not alone. The range of preparation activities that you will need to consider should include:

1. A materiality impact assessment through screening.
2. Setting metrics and targets for performance tracking.
3. Having a strategy that manages climate-related risks and opportunities.
4. Having plans in place for your people and capabilities to support mitigation action.
5. Performing measurements on a consistent basis using appropriate calculation methodologies.
6. Creating a baseline against which to measure change.
7. Producing both quantitative (for example, product ratings) and qualitative output (for example, tonnes of CO2e saved or carbon intensity scores for products).
8. Being able to report against the IT-related scope 3 emissions in your value chain.
9. Having appropriate verification processes in place.

---

[155] International Financial Reporting Standard, S2 - Climate-related Disclosures

[156] European Sustainability Reporting Standard E1 - Climate Change

10. Mapping mitigation plans to the broader business entity's climate transition plan.

If you needed any additional motivation to take action, you now have it. The aforementioned tasks may not be required to be completed tomorrow, but they are imminent. Regardless of the reporting standards, the preceding ten-point list constitutes a sensible action plan that encompasses the key outcomes required by your sustainable IT strategy.

Summary

The new reporting standards are yet another compelling reason for IT leaders to act. If there are three things that you should take away from the last 15 pages that will support your investment in creating a sustainable IT strategy, they are:

1. **IT is a material enabler** for many business prospects and will need to be reported. The new standards will require and regulate this.

2. **IT functions play a crucial role in managing sustainability and climate-related risks and opportunities**, both in terms of using technology to improve measurement and data accuracy and as a core business function that contributes to these risks and opportunities.

3. **Decarbonising digital is a key 'Decarbonisation Lever'.** The new reporting standards will require organisations to pull this lever both internally and externally throughout their value chain. They will be expected to disclose their performance in relation to set metrics and targets.

Decarbonising digital can no longer be seen as a choice, a wish-list item, or an altruistic exercise to be completed at our convenience. The new reporting standards and regulations are the first formal step to making this a 'must-do' activity and one that needs to be fully embedded into your normal operations.

Cutting Carbon, Cutting Costs: A Win-Win Outcome

Deloitte's research[157] reveals that there are four main factors driving businesses to address climate change:

1. Enhancing a company's reputation and obtaining the trust of their customers and clients.

2. The opportunity to save costs by taking action on climate change.

3. The need to comply with current and/or upcoming regulations.

4. Adaptability to the requests of clients and consumers, who are increasingly aware of sustainability issues and therefore demand more climate-friendly products and services.

So far in the book, I've covered items 1, 3, and 4 quite extensively, but item 2 relating to cost and savings needs addressing. It really is one of my least favourite topics, as it seems so shortsighted to talk about money when trying to address the biggest challenge that humanity has ever faced. But let's get real; the cash matters. The chart on the next page shows the results of the analysis conducted by Deloitte, and it is evident that the idea of cost reduction in tandem with climate action is seen as a significant motivator across all sectors.

I have witnessed these motivations first hand. Across nearly all of GoCodeGreen's client engagements (and I'm trying to remember the few that weren't), it becomes very clear very early on that businesses are driven by the opportunity for cost savings and find it challenging to align the

---

[157] Deloitte Insights, 'Cut carbon, cut costs', Feb. 2022

priority or investment to real climate action without having an associated cost reduction story to tell.

Climate action is needed to ensure reputation and trust but can also cut costs (top 4 drivers shown)

Question asked: What are the major factors driving climate action in your organisation?

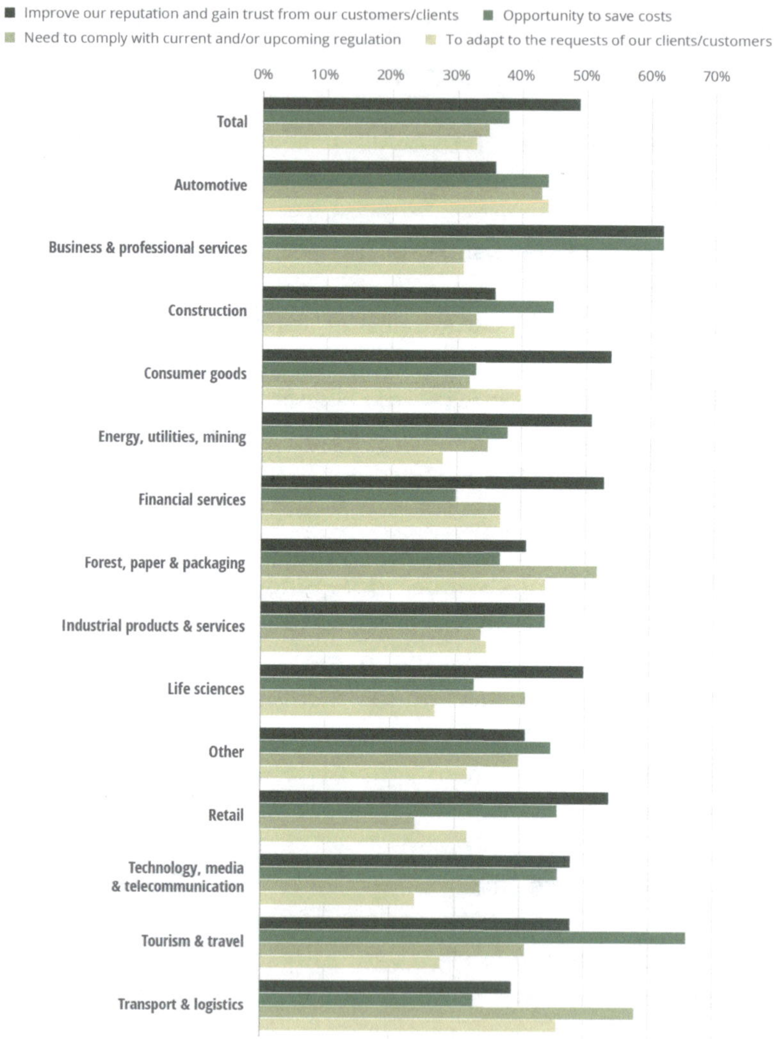

Source: Deloitte European CFO Survey, Autumn 2021.

Deloitte Insights | deloitte.com/insights

This need to constantly reconcile cost and climate always feels like a disappointment to me, but as I mentioned previously, we must also be realistic, as the majority of businesses face significant cost pressures as a result of the COVID-19 pandemic, global economic challenges, inflation, and the price of energy. I've learned that it is more powerful to support clients by helping them build a positive investment narrative that effectively interlocks carbon reduction with cost reduction. Over the rest of this section, I'll explain the approach that I've created to achieve this, and you can decide whether or not it can work for you in your organisation. I call it a 'Dual Outcome Investment Approach'.

A 'Dual-Outcome Investment Approach'

The good news is that it is possible to reduce both carbon emissions and costs. We are all aware that combating climate change is critical, but from a business perspective, the transition from idea to action is made easier when the investment needed to take action also generates cost savings. I refer to this as a 'Dual-Outcome Investment Approach' and use the following example to help bring it to life:

With the dual-outcome investment approach, a 10 percent allocation of existing IT operating costs and a 12 percent allocation of change investment can result in a 10 to 15 percent annualised reduction in IT run costs and a 25 to 35 percent reduction in carbon emissions.

The principle of conditional probability

How is this achieved? In simple terms, the Dual-Outcome Investment Approach ties investment to initiatives that drive both cost and carbon reduction in order to help leaders gain support for their sustainable IT strategies. The approach is loosely founded on the principle of conditional probability[158], in which the probability of an outcome depends on the occurrence of a preceding event or outcome.

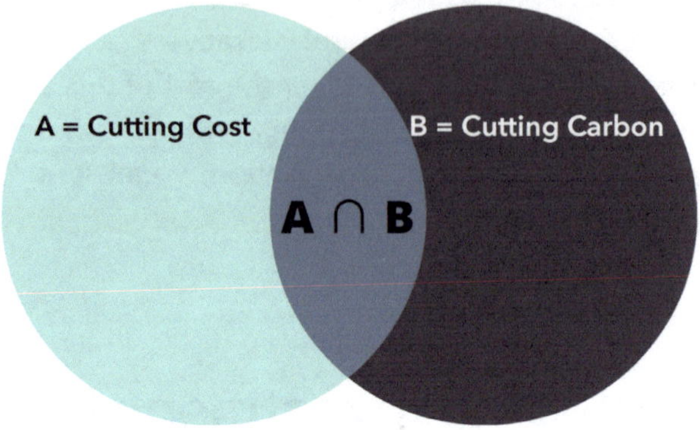

Consider that A in the diagram above represents your current cost-cutting initiatives, while B represents the actions required to decarbonise. The intersection area labelled AnB represents the connected activities that produce not only a dual-outcome but also a more likely outcome due to the activities' relationship. Based on this approach, prioritising activities that lead to AnB outcomes is a good way of utilising both existing investments and the momentum behind existing initiatives to mobilise your sustainable IT strategy.

As a first step, consider identifying and targeting AnB-type activities under the four categories listed below:

1. Reduce the energy demands of IT

This covers any initiative that reduces both the cost and energy consumption of ICT. I'm sure many of these activities will be familiar to you and may already be underway. For example:

---

158 Reference: britannica.com, 'Conditional Probability - mathematics', Stephen Eldridge, Sept. 2023

- Rationalisation of applications and removal of legacy software and infrastructure.
- Redesigning applications to increase their energy efficiency and cost to operate.
- Developing new applications with reduced infrastructure and energy requirements.
- Adjusting service levels and operational demands (e.g., reducing disaster recovery times), thereby reducing energy, infrastructure, and cost overheads for maintaining uptime demands that may not be required by customers.

2. Extend the lifespan of infrastructure components

It covers any activity that extends the life of infrastructure components, whether they are end-user, desktop, network, or data centre-related, and will result in longer use, lower replacement costs, and an extension of the life span of the carbon embodied in those components. Examples of activities include:

- Delaying the IT infrastructure refresh cycle by a further 12 months (or longer) by implementing sustainable software engineering and green design techniques to make more efficient use of existing IT resources.
- Adopting circular economy principles, recycling IT components, and reusing infrastructure as opposed to replacing them.

3. Use existing IT resources more efficiently

Covers a series of interconnected cost and carbon reduction initiatives that promote IT resource efficiency, for example:

- Utilising compute resources more efficiently in production environments.
- Reduce the number of environments for development and testing.
- Improving FinOps practices to maximise cloud computing's efficiency.
- Increasing DevOps maturity leads to improved speed of delivery, more efficient use of resources, and increased output.

- Implementing application modernisation and sustainable software engineering in order to produce more efficient software and optimise the utilisation of existing computing resources.
- Adopting a cloud-native software architecture to take advantage of on-demand and elastic computing capabilities.
- Implementing data rationalisation strategies to reduce data storage and retention needs.
- IT resources can also refer to people, so workforce decisions, such as using crowdsourcing and decisions to utilise IT resources in geographic locations with reduced grid carbon intensity to lower emissions.

4. Manage the IT supply chain

The impact of scope 3 emissions coming from the supply chain is a primary source of carbon costs. IT procurement and supply chain teams can play an important role in achieving cost efficiency and carbon reductions with the following types of initiatives:

- IT vendor assessments are based on their ability to meet energy efficiency and emission reduction targets at both the company and product or service level.
- Setting consistent standards and undertaking quality measurement across all scope 3 emission sources.
- Seeking visibility into multi-year decarbonisation strategies as part of the vendor evaluation process.
- Incorporating product and service level sustainability scoring into all pricing and information requests.
- Prioritising working with B Corp-certified companies committed to the triple bottom line of people, planet, and profit.

These activities would allow organisations to take advantage of their purchasing power and exert positive pressure on their supply chain to take action and decarbonise. They are low-investment initiatives that can be implemented swiftly and generate cost and carbon reductions.

A Modelled Scenario

Let's try to bring this to life using a modelled scenario where cost and carbon reduction actions are based on a 'typical' expenditure profile distributed across the categories of 'running' and 'changing' IT. Here, the Dual-Outcome approach allocates spend from existing costs and investment funding and targets the delivery of dual-outcome initiatives to achieve cost and carbon reductions per the table on the next page and description below.

In this scenario:

- 10 percent [B] of your current annualised IT operating expenses will yield a 5 percent [C] annual cost reduction.
- Investing 12 percent [E] of your current IT change costs will deliver efficiencies that reduce annual operating expenses by 10 percent [F].

When expenditure is aligned with the prioritised dual-outcome initiatives, the levels of cost savings and carbon reduction can be targeted, as shown in the table. The dual impact of spending 10 percent [B] and 12 percent [E] would be expected to materialise as follows:

- Greater energy efficiency in the operation of IT, with the potential for first-year savings of up to 35 percent (and additional savings in subsequent years through sustained investment).
- Up to a 25 percent annualised reduction in carbon emissions as a result of implementing IT transformation projects that also have energy and carbon reduction associated with their outcomes.

This is very clearly a modelled scenario, and you can modify the numbers, ratios, and reduction initiatives to better correlate with the context of your own organisation, but it shows how using the 'Dual-Outcome' approach can help construct a strong case for using existing funding to begin taking action.

COST OF IT (£)

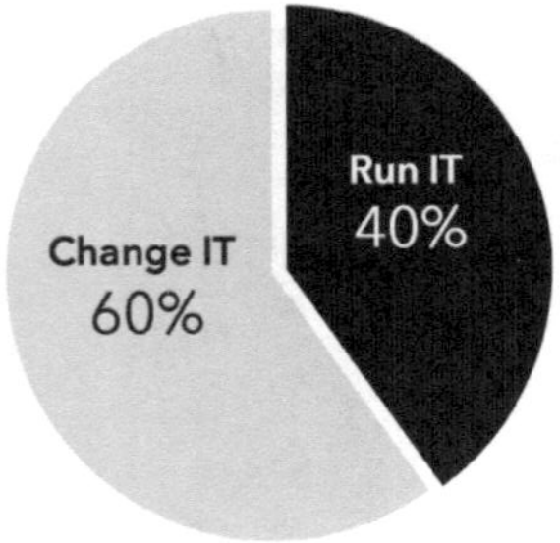

Run IT

Impact Category	[A] Associated Cost : % Share	Typical Cost & Carbon Reduction Levers	Investment Estimate (Share of [A])	In Year Saving
Infrastructure / Compute	50%	Measurement / IT Refresh choices / Extend life / Utilisation Improvements / FinOps	15%	-5% [C}
License / Maintenance	20%	Procurement / Supply Chain Targets	5%	
People Related	30%	Application Rationalisation / Travel / Service Levels / Automation / Location	5%	
		Average % Investment to Drive Efficiencies	**10% [B]**	

Change IT

Impact Category	[D] Associated Cost : % Share	Typical Reduction Levers	Investment Estimate (Share of [D])	Save on Run
Infrastructure / Compute	20%	Measurement / Circular IT / Modernisation / Cloud & Green Hosting / IT Choices	15%	-10% [F}
License / Maintenance	30%	Procurement / Supply Chain Targets / Vendor Reductions	5%	
People Related	50%	Engineering / Design / DevOps / Training / Travel / Location	15%	
		Average % Investment to Drive Efficiencies	**12% [E]**	

CARBON IMPACT OF IT (TCO2e)

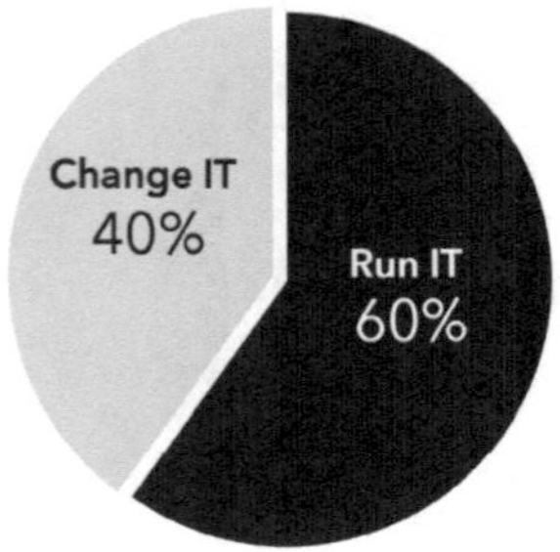

Run IT

Impact Category	Associated Carbon Cost : % Share	Reduction Opportunity
Infrastructure / Compute	50%	-35% [G]
Desktop Managed Services	20%	
Network Managed Services	15%	
People Related	15%	

Change IT

Impact Category	Associated Carbon Cost : % Share	Reduction Opportunity
Infrastructure / Compute	35%	-25% [H]
Desktop Managed Services	10%	
Network Managed Services	5%	
People Related	50%	

In the dual outcome model, an allocation of 10% of existing run costs and 12% of change investment can generate a 10%-15% annualised cost reduction to the ongoing cost of running IT and a 25%-35% carbon reduction.

Carbon reduction reality check:

In the last set of digital product assessments completed by GoCodeGreen during the first half of 2023, the total aggregated carbon reduction averaged 56 percent. This level of reduction opportunity aligns favourably with the figures modelled above and suggests that the model is conservative.

Conclusion to Part Four

The IPCC report has identified cost-effective measures for reducing greenhouse gas (GHG) emissions, intensifying carbon reduction, and enhancing resilience. Alongside the approaches covered in this book, we are not lacking solutions to the problem. And despite the rapidly closing window for tackling the climate crisis, a safe, habitable future is still possible. The risks of inaction are enormous, and the path forward calls for profound change. If we act quickly, it is still possible to limit the global temperature rise to 1.5°C. The world must reach peak greenhouse gas emissions by 2025, nearly halve emissions by 2030, and achieve Net-Zero CO2 emissions by the middle of the century. A 'all-hands-on-deck' approach is needed to ensure that the communities experiencing the most harmful impacts have the means to adapt to this new world. Governments, the private sector, civil society, and individuals must all take action to ensure the safety of our future.

Leaders across technology must prioritise action and use their influence to positively contribute to these changes. We must reduce our energy consumption and boost our energy efficiency in every aspect of our operations, because in the majority of GHG emission pathways proposed by the IPCC to limit global warming to 1.5°C, energy efficiency and a reduction in energy use are vital.

As technology has become increasingly indispensable in our modern world, the sustainability of IT is now crucial. According to a report by the United Nations Environment Programme, in the next decade, 70 percent of new economic value will be created through digitally enabled platforms. Technology leaders must balance this growth by making digital as sustainable and energy-efficient as possible.

We now know why it is so challenging to assess a digital carbon footprint. The complexities in the ICT value chain are extensive, but the tools are now available, presenting a real opportunity to accelerate decarbonisation. By doing so, we will limit our impacts on the planet's natural ecosystems. And technologists must help preserve nature's balance by saving every kilogramme of carbon dioxide possible.

Sustainable IT is an approach that prioritises the environment in the design, build, use, and disposal of computer hardware and software applications and business processes. It consists of software, data centres and cloud infrastructure, storage systems, networks, communication systems, and end-user devices. By adopting a sustainable IT strategy, businesses can increase their resilience by mitigating ESG risks, enhancing their reputation, and boosting their financial performance.

A significant barrier for organisations implementing sustainable IT practices remains the lack of understanding of the environmental impact of the digital world. 43 percent of technology executives are not aware of their company's IT-related carbon footprint. IT leaders will need to quickly improve as new reporting standards and regulations have now arrived. Noncompliance will lead to sanctions, but the cost of noncompliance will be substantially higher. Customers will evaluate prospective vendors through the lens of CSRD, and consumers will make choices based on how their purchases affect the environment. Compliance with CSRD can improve brand trust and image, and businesses can build employee, client, and public trust if sustainability is prioritised.

Reporting standards such as IFRS S1 and ESRS E1 mandate the inclusion of sustainability-related disclosures and financial statements in the same report. Although there is currently no formal guidance on how IT teams should interpret and implement these standards, it is now inevitable that IT activity-level reporting will be required in the very near future, and as soon as 2024.

A 'Dual-Outcome Investment Approach' is a good way for IT leaders wanting to mobilise their sustainable IT strategy, in which a percentage of existing run costs and change investment in ICT are allocated to decarbonising digital, resulting in both an annualised reduction in the cost of operating IT and a reduction in carbon emissions. This win-win outcome can help us move from observing the problem to real action.

So much of what we need to take action is now available. As a reminder, look at the diagram below. Refer back to each section as highlighted, and combine and tailor the guidance to your organisation to help you create an overall framework for decarbonising digital.

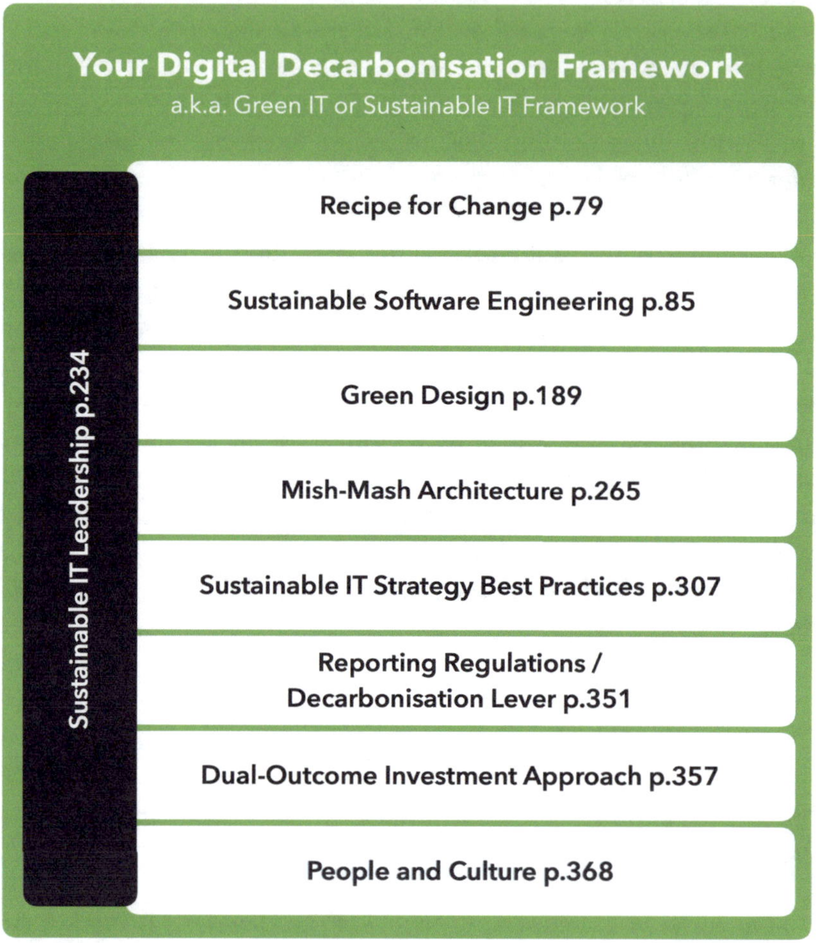

A determination and commitment across ICT and digital businesses to make this happen is now required. This will also need the people and culture within organisations to change, and that will be covered in the final chapter of the book.

Part Five: The Power of Us

People and Culture

Employees are the most valuable asset in any business. It's the people who will be responsible for carrying out the work and making things happen. So creating a vision, establishing objectives, building momentum, and assembling a team that can make sustainable IT a reality is not only vital but will require exceptional leadership. In order to put these goals at the forefront of an organisation's mission, many companies are now choosing to set and monitor progress using the triple bottom line rather than the conventional bottom line.

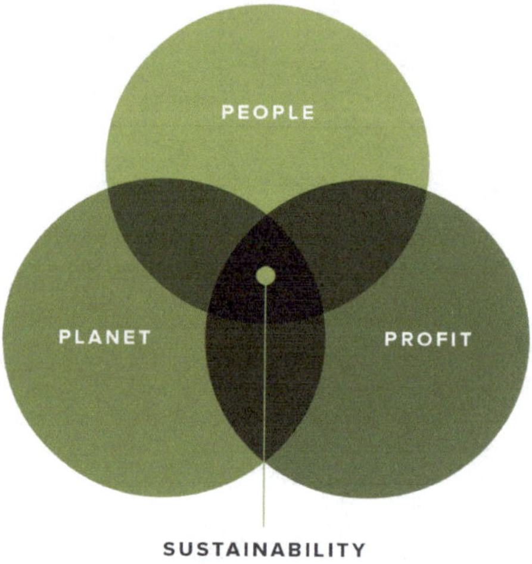

As shown in the diagram above, people, planet, and profit (also referred to as the 'Three Ps') form the basis of the triple bottom line. By keeping these three factors in mind, a business can evaluate its sustainability efforts and ensure that profit does not overshadow them. But let's be clear: sustainability also provides ample opportunities for profit. Once considered separate goals, many businesses now find that employing sustainable practices saves them money and attracts new, sustainably-minded customers who are interested in supporting businesses that contribute to environmental and social good. As an IT leader, it is always good to remind yourself that all of this is only made possible by your people.

GoCodeGreen is a certified B Corporation, so I am completely aware of the benefits of the triple bottom line and what it takes to achieve them. I believe that creating organisations that value and invest in people in order to foster a culture with sustainability at its heart is a goal we should all pursue. I won't pretend it was an easy process; ours wasn't and required huge effort and commitment, as it should. Different organisations will take different paths over different timescales to get there, but the objective is something we should all strive to achieve.

"You don't have to leave your values at the door when you come to work."

Harvard Business School[159]

I often wonder why individuals who are so passionate about climate action in their personal lives appear to lose their enthusiasm when they enter the workplace. Is it because their organisations care less than they do? That there is a lack of recognition and incentives for climate-related work? Or merely exhaustion from attempting to be heard? Whatever it is, we must do better as leaders. To create an environment in which people feel empowered and valued for their desire to make a difference.

In the following sections, I will discuss how to structure an organisation to effectively deliver sustainability in IT, as well as the importance of knowledge and skill development to maximise the capacity for rapid action. This is a crucial chapter in the book.

People and culture really matter when it comes to climate action. Your people are your biggest lever for change.

---

[159] Harvard Business School, 'Sustainable Business Strategy', HBS Professor Rebecca Henderson, (undated)

People

Organisational structure

According to a 2021 McKinsey report[160], very few businesses have adopted organisational structures designed for sustainability as a material business outcome. As a result, activity is often short-term in nature and task-based. McKinsey recommends focusing less on task-level activity and more on broader, measurable outcomes.

"Success is more likely when executives empower the organisation to engage proactively and strategically hold them responsible for creating measurable impact. Only then will companies be able to maximise the value at stake from their sustainability initiatives."

McKinsey

But to enable 'measurable impact' to drive a sustainability agenda, the vast majority of organisations will need to rethink elements of their organisational design. This will help to recalibrate business trade-offs, decision-making, and governance processes, as well as employee and leader attitudes. The four quadrants in the diagram on the left can help create a foundation for

160 McKinsey Sustainability, 'Organising for sustainability success: Where, and how, leaders can start', Aug. 2021

challenging your existing organisational structure and determining how to best implement changes that will promote sustainability, attract support, and empower people to act. The four main focus areas are:

1. Design the organisation around material topics

Start by defining a list of sustainability topics that are important to the IT organisation; for example, a measurement-based assessment is a useful starting point for developing a backlog of actions that can be prioritised. This will support decision-making and identify items that will have the largest impact on agreed-upon targets. In practice, large central structures aren't the best at addressing these priorities.

Even if there is a centre of excellence dedicated to a specific topic (such as executing a development plan for sustainable software engineering or accelerating migration to the clean cloud), it does not necessarily need to be part of a central team. Instead, it is better to embed them within a functional unit that has specialised knowledge on the topic or will be primarily responsible for leading the IT department's response to it.

2. Empower the Sustainable IT team to make decisions

From experience, the most effective approach is to have a smaller Sustainable IT Centre of Excellence (CoE) with dedicated resources in the appropriate functional divisions responsible for executing the detailed planning and implementation of Sustainable IT. A federated responsibility model with lightweight but effective coordination to drive consistency in approach and data is the best way to empower teams to take action.

Establish a lean CoE for sustainable IT whose mandate is to incubate new ideas and integrate sustainability-based initiatives across the organisation, help establish targets, measure performance, and act as an enabler for the entire business. Empowering this CoE with decision-making authority to implement change, especially high-priority material topics that affect multiple functions, is a highly effective approach and increases the likelihood of success.

Sustainability is not just a compliance issue for the majority of businesses but rather a strategic and operational one. There is no one structure that is ideal for every business, and even within the same business unit, the structure may vary to be most effective. A well-designed organisation can help harness the capabilities required to capture value and manage the risks associated with achieving sustainable IT in a systematic and even transformative manner.

3. Create an organisation structure based on collaboration

Collaboration is essential, as no one individual or team can deliver a sustainable IT strategy for an organisation. Designing an organisational structure that is incorporated into and compatible with the larger business context will significantly improve the likelihood of success. As a consequence of the interconnected activities and decisions required to drive action, sustainability necessitates a multidisciplinary approach. These complexities are especially evident in IT and digital functions, where the stakeholder groups come from across the entire organisation.

I use the table on the next page as a starting point to illustrate the variety of stakeholders and focus areas that are needed to make a sustainable IT strategy successful. It can be considered an organisational framework where each organisational group must take the lead in distinct areas of expertise and responsibility; however, success is only achieved by bringing everyone together to enable deep team collaboration. This 'all hands on deck' approach begins with leadership consensus on the purpose and value of investing in a sustainable IT strategy, and how well it aligns with and accelerates the organisation's broader sustainability commitments. Ideally, the framework is established before the actual 'doing' begins, as this creates the best foundation for accelerated action.

	Digital & Technology / IT Leaders	Product Management & Transformation Leaders	ESG & Compliance Leaders
Are you ALL in?	**"All hands on deck now!"** - Intergovernmental Panel on Climate Change (IPCC), Feb 2022 **Are all three leader groups agreed on the need to act and the value of sustainable IT? For success, it is critical that you are all fully aligned on purpose; the opportunity and need to reduce digital related carbon emissions; and support developing your strategy and targets for decarbonisation.**		
Next Best Action to Deliver the Proof of Value	Action: Drive the value chain assessments by identifying applications and allocating resources to support the PoV. Outcome: Baseline comparisons and data to set scale of carbon reduction opportunities across digital and IT estate.	Action: Build the customer value proposition for your decarbonised digital products and services to support the PoV. Outcome: Clear ratings and decarbonisation insights that will help shape the product backlog and future KPIs / OKRs.	Action: Use the PoV to build awareness that Digital and IT are key contributors to ESG commitments and achieving Net Zero. Outcome: Clearly mapped PoV results to GHG Protocol ICT Sector Guidance and alignment to Net Zero commitments.
Building Success & Contributing Toward Your ESG and Net Zero Commitments	**Mission.** Build the storyboard and mission. Technology can be a a pioneer and major contributor to decarbonisation and helping to achieve your organisations ESG commitments and Net Zero ambitions. Baseline, Set Targets and Act.	**Investment.** Build the 3-5 year decarbonisation investment case. Establish clear a clear OKR at top level and specific carbon reduction ROI metrics and targets. Actively support delivery of IT enablers in PI planning.	**Targets.** Help guide and shape digital decarbonisation metrics with an ESG lens. Support building the 3-5 year sustainable IT strategy to align with organisation wide ESG targets and commitments toward Net Zero.
	Capability & Culture. Build the digital decarbonisation execution capability and mission into the culture of your technology teams. Boost morale. Create education opportunities (e.g. sustainable software engineering) for your people. Use sustainable IT to attract and retain talent.	**Planning & Prioritisation.** Based on mission, strategy, roadmap and investment - support digital product development backlog prioritisation to achieve key OKR and KPI targets. Work with IT to create a multi-year plan with near, medium and longer term carbon reduction targets.	**Carbon Accounting.** Help provide carbon accounting support and advisory to IT and Product Management. Create audit and tracking for OKRs and KPIs. Support deduplication of any Scope 1, 2 and 3 measurements in reporting.
	Strategy. Map decarbonisation actions to your strategy. You will be doing things already - now measure the carbon impact (e.g. Cloud First, Modernisation, Simplification). Accelerate your existing strategy (e.g. Legacy removal). Introduce new decarbonisation levers.	**Process Simplification.** Actively seek to identify obsolete, redundant and unnecessary business processes to support rationalisation and simplification of the product. Approve removal of unnecessary or over-engineered features.	**Carbon Neutrality & PAS 2060.** Oversee the pathway, tracking, and documentation audit to help achieve Carbon Neutrality for digital products and services. Provide consistent and clear decision making frameworks for PAS 2060 certification.
	Supply Chain. Immediately introduce carbon assessments into supply chain and procurement decisioning. Make smart carbon decisions now, stop adding to the challenge, and use your vendors to take action.	**Customer Needs.** Review service levels and customer needs to help simplify and remove complexity from processing and IT operations.	**Advisory & Independence.** Align ESG team members to supporting the PoV and provide guidance to the project teams to ensure connectivity with ESG strategy and approaches. Help create a 'one team' approach. Own the GCG relationship.

4. Prioritise around tools, measurement, governance and operationalising action

Ultimately, everything should be oriented towards the creation of an organisational structure that facilitates, enables, and accelerates action. Once the material reduction opportunities are understood and prioritised, the structure is in place to facilitate decision-making, and a collaborative framework is established, the final ingredient is equipping people with the knowledge, tools, and processes necessary to be successful. Connect all of this back to the measurement tools, approaches, and best practice steps for building a sustainable IT strategy, and you will have started to create a framework for taking effective action.

Building the capability to take action

People are essential, but what if they don't have the skills or knowledge to solve the challenges or work at the required pace? This book has already covered the software engineering, design, and leadership knowledge needed to bring about the scale of change required. IT leaders should consider the following eight areas of capability that are needed and invest accordingly.

1. Everyone should be given basic knowledge

Ensure that your employees have a basic knowledge of environmental science and climate change before implementing your sustainable IT strategy. There is no need for advanced science, but it never ceases to amaze me that so many people are unable to explain the fundamental principles behind global warming and climate change. Better understanding the problem you are trying to fix always helps.

If you have read this book from the beginning, then you will have gained the basic knowledge already. You will also understand why this is not someone else's problem and why one-dimensional perspectives such as 'everything will be fine if we move to the cloud' or 'once everything is running on renewables, we can consume as much as we want' are overly simplistic and simply not in line with the deep-rooted changes needed to decarbonise digital on a sustainable, long-term basis.

It is essential for leaders to invest in the education of their employees, not only in their role-based learning but also in understanding the broader challenges of the industry, environment, and community in which they operate, as well as the systemic connections that exist between them. Understanding all of this helps contextualise the effort that is needed. We are about to change decades of behaviours and choices in order to affect the scale of change that is now required.

2. Understanding the business response to decarbonisation & sustainability

Understanding how your business performs in relation to the metrics and targets set for digital decarbonisation and sustainable IT will soon become a regular topic of discussion within your organisation. How would you respond to the following questions? What is your company's quantity of IT-related carbon emissions? Does your ICT supply chain undergo a comprehensive impact analysis? Are you aware of the environmental impact of the digital products and services you provide to customers? Are you prepared for the new reporting standards and regulations? And how do the answers to any of the above fit into the broader organisation's commitments to achieving Net-Zero? In fact, the bigger question you will need to prepare for will be, 'What will IT contribute to our organisational Net-Zero commitments? And when?'

To buy into your sustainable IT strategy, you will need to start building a response to these questions and help your business and people understand how the answers will form a cohesive plan of action. Creating an accurate picture of your current state is one of the most effective ways to start this process and will provide a baseline from which you can make progress.

3. Leadership at all levels

It will require more than one leader to drive a sustainable IT strategy, so building strong leadership capabilities across teams will be key to success. An organisation will require leadership qualities that include setting a vision for sustainability in IT, then turning that vision into a reality by articulating clear goals, defining a strategic plan for achieving those goals, and equipping and empowering each team member to implement the plan.

Tackling climate change can be an emotionally charged topic based on the individual beliefs of team members; therefore, increasing emotional intelligence, the ability to communicate change effectively, and maintaining authenticity are leadership traits that will be required as you embark on this journey. It is best to invest in developing the skills and knowledge across all leadership levels to think quickly and critically, consider all perspectives, and make decisions with people, planet, and profit in mind.

Sustainability and addressing climate change within a business context are frequently seen as 'side of desk' activities; however, to achieve everything we've discussed in this book to enable a deep decarbonisation of digital, you will need to raise the stakes. This needs to be at the forefront of everything you do.

4. Thinking with a sustainability mindset

Individuals and teams with a good knowledge of sustainable business practices are better equipped to identify opportunities, implement impactful changes, and contribute positively to sustainability efforts for their organisation. With a greater understanding of the cause and effect of various technical decisions, teams will be better placed to identify opportunities to reduce energy consumption and lower carbon impact.

Educating employees on the significance of sustainability and the initiatives you plan to implement will require an investment of time and resources, but it can help create a company culture centred on sustainability, which will have a positive and scaled impact in the future. As environmental issues and the understanding of ICT's impact evolve, you can help prepare your organisation's readiness and ability to adapt and act.

5. Innovation and problem-solving skills

To address the sustainability of IT, it will be necessary for people who are adept at problem-solving and developing innovative answers to new and complex challenges to be part of the challenge. Many organisations already struggle to incorporate innovation into their work practices and will need to double their efforts if they are to create an environment in which teams feel encouraged to come up with innovative ideas to help address the sustainability of IT. Leaders must make their organisations ready to experiment and embrace innovation.

Investing in and building a culture of innovation in which products, processes, and the workforce are continuously improved to create a more sustainable digital future is crucial if we are to be successful. The status quo will not work. Experimentation, leveraging the broader ClimateTech ecosystem, and a willingness to work with less-than-perfect data are all techniques that can help accelerate action and introduce new approaches and solutions within an organisation.

6. Being confident with data

Taking action to improve the sustainability of IT requires data to help determine where and what to prioritise, as well as how to monitor and measure performance. Teams will need to collect, analyse, track, and report on both climate and IT-related data. The ability to work with varying levels of data quality, coverage, and accuracy will also be a test for all those involved. There will inevitably be varying interpretations of data requirements, so teams need to avoid spending too much time seeking perfection in order to drive analysis and decision-making. Selecting the best measurement tools available to promote consistency and recognising that data will change and improve over time should enable more pragmatic decisions that won't delay mobilisation and action.

7. Communication

There are two aspects of communication that need to be considered: the first is 'what' is being communicated and why it is significant; being clear on purpose and outcomes encourages engagement; and the second is 'how' to communicate and the effectiveness of messaging.

The 'what'

Organisations that make an effort to define their climate change purpose and use this as the basis for their sustainable IT strategies will engage their employees more effectively. Those who focus immediately on technical activities, such as green coding, without first establishing a context for 'why' such activities are important and a framework for achieving them frequently have to reset later on. Aim to build a strong sense of purpose and partner this with a structure that allows employees to see how their contributions will impact the organisation and make a difference.

The 'how'

Using effective communication techniques to support leaders and teams to communicate the 'what' and 'why' can help to develop a deeply ingrained culture of sustainable action and thinking within an organisation. I've found that the following communication techniques and approaches can be highly effective:

- Sharing data and statistics that communicate the urgency of sustainability and climate action and doing so through a digital and IT lens to make it relevant to technologists works exceptionally well, particularly when mobilising or gaining initial buy-in from employees.
- Use data visualisation techniques to present compelling data trends and insights in a way that IT professionals find visually appealing and understandable. Be honest and avoid marketing materials or information that could be interpreted as greenwashing.
- Storytelling that illustrates why action is so important and identifying external experts who can contribute different points of view and inspire employees to engage and want to participate in taking action.
- Encourage personal reflections and ask individuals questions that elicit thought about their relationship to a sustainable digital future, for example, "How do I want to contribute to the world that our children will inherit?" or "What role can I play in redefining the future?" are the types of questions people are asking themselves in their personal lives; giving them the opportunity to take action in their professional lives can positively reinforce how they feel about the organisation they work for.

8. Understanding Sustainable IT

Building a knowledge base covering sustainable IT to standardise understanding and rally everyone behind the opportunity to take action can help make the execution of your sustainable IT strategy more effective. Sustainable IT knowledge will help you design, develop, and maintain the digital products and services of the future. This encompasses everything from sustainable software engineering, green design-thinking, and sustainable IT leadership skills to the frameworks and models needed to deliver the decarbonisation of digital. Sustainable IT is not just about being aware of the environment; it also involves understanding how technology can be used to improve our lives while protecting the planet.

It can help us do things differently, as well as help us do things better.

Culture

The desire to be sustainable is gaining importance among the workforce. Companies that prioritise sustainable IT will attract and retain talented employees who share their environmental values, and establishing a culture for executing your sustainable IT strategy will be a major enabler to unlocking the power of your people. But as you develop this culture, carefully consider the following topics:

Don't ignore eco-anxiety

It is important to recognise how large and frightening climate change is and feels. According to a study published in The Lancet[161], more than 50 percent of 16 to 25 year-olds worldwide reported feeling sad, anxious, furious, powerless, helpless, and guilty about climate change. In the same article, it was noted that more than 45 percent of respondents replied that these feelings negatively impact their daily functioning.

"We are undergoing the biggest shift humanity has ever gone through, we don't ever talk about how to emotionally deal with that."

The Climate Optimist [162]

---

161 The Lancet - Planetary Health, 'Climate anxiety in children and young people and their beliefs about government responses to climate change: a global survey', Dec. 2021

162 TheClimateOptimist.com, Anne Therese Gennari, Founder

And it is not just the next generation of employees that feel things need to change; it is also the existing workforce. Your teams need to feel engaged and part of a process that translates all the talk and good intentions into actions and decisions that make them feel like they are part of the solution to climate change.

Employee demand

Companies that demonstrate genuine concern for environmental, social, and governance (ESG) issues are better able to attract and retain employees in a time of widespread talent shortages. The importance of environmental sustainability among customers and employees has increased by over 51 percent, according to a study by IBM's Institute for Business Value[163].

The sustainability credentials of an organisation directly influence the positions that candidates accept or prioritise. According to the same report, two out of three respondents are more likely to apply for and accept a job (67 percent) if they believe the company is environmentally sustainable, and this percentage rises to 80 percent for businesses that prioritise the planet and people.

"Only 21 percent of survey respondents believed their current employers are sustainable. In a competitive talent market, that's a big risk."

IBM Institute for Business Value

According to the study, one-third of those who left their positions in 2022 transferred their skills to a company that they viewed as more environmentally sustainable (35 percent) or socially responsible (40 percent). One-third of these job changers took on positions that allowed them to

---

163 IBM Institute for Business Value, 'Balancing Sustainability and Profitability: Consumer research highlights how businesses can protect people, planet and the bottom line', April 2022

influence and contribute towards environmentally sustainable outcomes. In addition, roughly the same proportion of those who accepted a new job in the past year accepted a lower salary in exchange for the opportunity to work for a sustainable or socially responsible organisation.

Leading from the Front

Pioneering leaders will demonstrate their commitment to sustainability by tying their compensation to their performance in this area. This is crucial so that the rest of the organisation can buy into the change that leadership is attempting to implement.

To achieve success in sustainability, leaders must be accountable for their actions. Linking executive compensation to performance against sustainability targets will become the norm. By choosing to do this, leaders can demonstrate their commitment to advancing their organisation's sustainability efforts, and a number of big corporations are already starting to introduce these changes, for example:

NEWS ALERT: April 19, 2022 (Reuters): Payment processor Mastercard Inc. (MA.N) will link all employee bonuses to environmental, social, and corporate governance (ESG) initiatives, expanding an earlier program that was limited to its senior executives, Chief Executive Michael Miebach said on Tuesday. The move will help Mastercard achieve its goals of cutting carbon usage, improving financial inclusion, and achieving gender pay parity. Mastercard in November accelerated its Net-Zero timeline by a decade, to 2040 from 2050.

Achieving sustainability doesn't just happen, and as I often say, there is no magic switch to flick. It is a significant organisational initiative and, as such, requires prioritisation, people's support, and effort. As previously stated by Capgemini, only 34 percent of businesses currently incorporate sustainable IT into their board-level agenda, and only 39 percent of these organisations have a governance body responsible for supervising sustainable IT. Only 50 percent of this latter group includes CIOs as members. Technology leaders cannot be silent any longer; we need to find our voices and lead from the front.

Building a Sustainable IT Culture

> **"Change employee behaviour and create the right culture, ensuring IT is used in a way that aligns with the organisation's wider sustainability goals"**
>
> Capgemini Invent[164]

There are four important steps technology leaders can take to accelerate cultural change within their organisations and create the right environment to support their sustainable IT strategies:

1. Understanding and team capability

I've previously spoken about the importance of knowledge, and IT leaders should consider how they assess the capability of their teams to take action, as this is one of their main levers for affecting change. Assessing your people can help identify areas of technical and management knowledge that can be enhanced in order to accelerate action. Throughout each part of this book, I have given you access to the self-assessments I created with the team at GoCodeGreen, covering sustainable software engineering, green design, and sustainable IT leadership. Consider using these at scale across your

---

[164] capgemini.com, 'Reinventing work and sustainability' (undated)

organisation to build a baseline view from which you can target improvement if needed.

GoCodeGreen has completed these people assessment exercises across a range of organisations from different industries. In fact, we now insist on assessing the people who work on a product alongside the software itself as part of our measurement process in order to provide clients with the most comprehensive view of how to take action. We found that, on average, 70 to 85 percent of respondents rated their own knowledge of sustainable software engineering, green design, or sustainable IT leadership as either 'poor' or 'fair'. These results were consistent across age groups, gender, location, and experience, indicating that most IT organisations require broad as well as targeted learning to be provided to their employees in order for them to be ready to take action and effect change.

On average, between 70 and 85 percent of respondents rated their own understanding of sustainable software engineering, green design, and sustainable IT leadership as 'poor' or 'fair'.

All technology leaders should consider the state and readiness of their people-based capabilities as a top priority and begin to implement plans to up-skill their teams if they are to successfully execute their sustainable IT strategies.

2. Creating a champion network

Creating a sustainable IT culture requires an organisation-wide focus on responsible technology use. Technology leaders should consider establishing a network of sustainability champions throughout their organisations. Champions can originate or be seconded by the company's Sustainable IT Centre of Excellence or by ESG team members with expertise in technology.

These champions can help to raise awareness, facilitate training and workshops, and generally motivate employees to take action; they should focus on using success stories to promote and deepen buy-in to achieving IT-related sustainability goals.

3. Collaborate and focus on the right things

Technology leaders must appeal to the entire organisation for support of green technology-based initiatives. A culture of collaboration is needed to support both top-down and bottom-up approaches to engagement across the organisation, where all employees can be involved in identifying areas where the company can improve its sustainability position.

Establishing clear objectives and targets and making capital available for projects that can drive the sustainable IT strategy forward according to agreed-upon objective key results (OKRs) and outcomes ensures that effort is focused on the right things. An element of caution is required, as the enthusiasm for sustainability can sometimes prompt people to rally around visible issues at the expense of critical sustainable IT issues. People can spend a great deal of time and effort implementing recycling for coffee cups and plastic bottles in the office or going out to plant a tree on a community day, but not enough time considering the server room and the digital products that run from it, which will have a significantly greater carbon footprint but are less obvious. Be conscious of how to balance effort and focus on initiatives that make the greatest impact.

4. Using sustainability to boost innovation

Resistance to change is one of the greatest obstacles technology leaders face when advocating for sustainability. People frequently believe that sustainable IT initiatives will require them to make difficult trade-offs, for example:

- They'll have less access to high-performing technology.
- There will be greater restrictions on how people execute projects that require substantial computing power.
- There will be lower productivity because access to high-performing technology will be increasingly limited.

Instead, we must find solutions that are simultaneously more effective and reduce our environmental footprint in order to dispel the notion that a trade-off is always necessary. Technology leaders have the opportunity to encourage innovative responses to overcome resistance to change, and when a trade-off is required, employees should be supported to experiment and be innovative in working around them.

Conclusion to Part Five

The triple bottom line, also known as the 'Three Ps' of People, Planet, and Profit, is a strategic approach that assists businesses in driving forward their sustainability efforts and preventing profit from overshadowing them. Sustainability actually offers numerous profit opportunities, as it can save businesses money and attract consumers interested in supporting companies that contribute to environmental and social good.

Creating a company that recognises and invests in the value of its employees is crucial for achieving the triple bottom line, and exceptional leadership is required to create a vision, establish objectives, and assemble a professional team capable of making sustainable IT a reality.

At the beginning of Part Four, I talked about the 'Silent Technology Leader Syndrome' (page 234). When changing the culture and engaging people, it is absolutely critical that IT leaders find a voice and use the 'what' and 'how' communication techniques described in this section.

As leaders, we have to create a culture where our employees feel empowered and valued for their desire to make a difference. And when preparing employees and constructing a culture to implement their sustainable IT strategy, technology leaders should focus on two key areas:

- Employee development to provide the skills and knowledge to maximise their capacity for taking action within their roles.
- Developing a culture that values sustainability and encourages employees to act through innovative problem solving.

By doing so, we are unlocking one of the most powerful resources we have available to take action on climate change: our people.

Practical - Sustainable IT for Technology Leaders Checklist

Use the Sustainable IT for Technology Leaders Checklist on the next page as a constant reminder of the actions required to implement your sustainable IT strategy.

You can also access and download the checklist using the QR code above, or alternatively, visit: www.gocode.green/sustainabilityassessmenttools

GoCodeGreen
Decarbonise Digital®

Sustainable IT for Technology Leaders Checklist

Fundamentals for Success		Tick all that apply
Leadership Understanding	Reflected on own understanding of the carbon and environmental impact of IT, reviewed access to data to support decisioning, and taken steps to deepen knowledge and methods to instil a sustainability led mindset across technology practices.	☐
Sustainable IT Knowledge across Management Team	Considered the capabilities, experience and level of understanding across your management team and improved their ability to decarbonise your IT estate.	☐
Data Availability and Readiness	Understood the data needed to drive accurate carbon impact measurement across technology and started to take steps to improve and automate data to support the embedding of sustainability into decisioning.	☐
Baselined Understanding	Measured and baselined the carbon impact across all core elements of the technology estate (including infrastructure and software) and had this independently verified.	☐
Cloud Performance	Understood the benefits and challenges involved in extracting maximum sustainability based opportunities from the cloud and making informed decisions and choices at application level.	☐
Investment Decisioning	Made and supports investment prioritisation and allocation of spend based on the carbon and environmental impact of technology choices.	☐
Metrics and Performance Targets	Implemented a clear set of consistent metrics and targets across the technology and application estate and is able to measure and track performance on a regular basis.	☐
Reporting and Obligations		Tick all that apply
Carbon Reporting	Understood current and forthcoming carbon reporting standards (CSRD/ESRS) and IT activity metric level requirements and is actively preparing for them. Understands difference between ESG reporting and carbon reduction action related measurement.	☐
Scope 3 Understanding	Built the data and insights to manage Technology and ICT related Scope 3 impacts in your value chain to better inform spend and use decisions.	☐
Readiness for IT Level Reporting	Prepared teams and data to provide IT activity metric level reporting to support new standards and guidelines, and actively engaged with ESG team to understand timeframes.	☐
Process and Tooling		Tick all that apply
DevSustainOps	Incorporated sustainability focused activities and assessment criteria into DevOps and FinOps processes using DevSustainOps or similar method.	☐
Design Stage Intervention	Incorporated design stage sustainability assessment and purposefully making solution, service, UX, UI and engineering decisions before coding commences based on potential carbon impact and energy usage. Actively ensuring new product development is low carbon intensity.	☐
Measurement	Able to measure, baseline and establish appropriate metrics and targets at both pan-ICT and software / platform level to support action to reduce carbon impact and improve energy efficiency.	☐
IT Procurement Decisioning	Implemented a standardised approach to assessing the carbon impact and energy efficiency of new software products and ICT components, and engaged vendors to responds as part of any purchasing decision.	☐
Sustainable IT Centre of Excellence	Considered and established as appropriate a centre of excellence (or similar) to coordinate and consistently enable data provisioning, data assessment, actioning and reporting across the IT estate. Has identified the importance and value of setting, measuring and tracking performance against sustainability led targets.	☐
People and Culture		Tick all that apply
Understanding your Capability	Measured the teams understanding of green coding and design techniques and identified areas where investment in learning and development is required.	☐
Learning Pathways	Mapped learning needs across team and created pathways across disciplines to develop their knowledge, capability and experience in Sustainable Software Engineering Practices and Techniques (Green Coding) and Green Design.	☐
Customer		Tick all that apply
Customer Journey / Value Chain / Client Goals	Engaged business stakeholders and mapped the digital architecture to business / customer value chains or goals, and able to articulate the importance of digital carbon reduction to support investment and transformational decisioning.	☐
Product / Service Differentiation	Understands the value of carbon impact reduction from a digital product or service perspective and embeds this thinking into business and customer engagements to harness support for action.	☐
Understanding Customer Needs	Understands the customer and investor demand for lower carbon digital services and importance of sustainable IT practices.	☐

Final Comments

The manga storyline is at an end, and your journey through the book is nearly complete. There has been a great deal covered; let me sum up and conclude with my final thoughts.

The IPCC report emphasises the importance of addressing climate change and creating a habitable future. It stresses the need for immediate action to limit global temperature rise to 1.5°C, with emissions needing to peak before 2025, nearly halve by 2030, and reach Net-Zero by mid-century. Technology leaders must prioritise action and utilise their influence to contribute to a global movement to tackle climate change. This isn't somebody else's problem.

As technology has a substantial carbon footprint, sustainable IT is crucial for businesses. Technology leaders must contribute to making digital as sustainable and energy-efficient as possible. Digital carbon impact is measured in terms of the greenhouse gas emissions generated by digital software, technology resources, devices, tools, and platforms. ICT and digital need to be part of the solution to climate change, and through this book, I've shown you a recipe for change and created the framework to take action. But we need to start now.

Sustainable software engineering is a growing trend that seeks to enhance application efficiency, reduce energy and space consumption, and reduce the long-term environmental impact of IT infrastructure. The sustainable software engineer is needed to develop digital solutions that are energy and carbon efficient. Utilising sustainable software engineering principles and the twelve key engineering techniques can help develop greener, more environmentally responsible software. These principles and techniques, as explained in Part Two, are available now and can be applied immediately.

To achieve a low-carbon, sustainable digital future, the importance of great design is paramount. Green design, which incorporates the principles and techniques described in Part Three, helps determine the right problems to solve prior to coding. Design-based thinking is the most effective way to incorporate 'green' into solutions, and the UK's Design Council's Double Diamond paradigm offers a method for concentrating on both problems and solutions in the design process.

Adopting a green design framework can help to maximise the efficiency of digital applications, lessen their carbon footprint, and reduce their negative impact on the environment and nature. It involves reducing electricity consumption, incorporating environmentally friendly elements, and choosing energy-efficient visual design features. The implementation of green design principles can improve performance, usability, accessibility, operational costs, employee morale, create mission-driven teams, and increase customer satisfaction.

Measuring a digital carbon footprint is difficult and complex, and many technology leaders remain unaware of IT's environmental impact. To overcome this, leaders must take immediate action to find tools to measure and reduce their digital carbon footprint. The need for measurement is now critical as new reporting standards and regulations have already been introduced. The best practices and framework set in Part Four will help IT leaders make a significant contribution to achieving their organisation's Net-Zero commitments.

The triple bottom line, also known as the 'Three Ps' of People, Planet, and Profit, is a strategic approach that assists businesses in driving forward their sustainability efforts and preventing profit from overshadowing them. To accomplish the triple bottom line, technology leaders should prioritise employee development and build a culture that prioritises sustainability and encourages employees to take action. As we concluded in Part Five, unlocking the power of our teams and people to take action should be a priority for all technology leaders.

"Measuring your impact takes time". Anyone who has undertaken the B Corporation assessment, verification, and certification process has seen these exact words when logging into the B Lab system. In a digital world where everyone is short on patience and demands immediate results, 'taking time' is typically viewed negatively. In reality, complex changes, such as organisational structures, culture, business processes, technology, and data, require time to understand and unravel. B Corporation has mastered the concept of value in their offering; we are willing to invest time and effort because we value the outcome. Please bear this in mind when deciding to invest the effort into decarbonising digital; it may take time, but the outcome will benefit us all.

Everything, everywhere, all at once

At the conclusion of its sixth assessment cycle, the recent Intergovernmental Panel on Climate Change (IPCC) synthesis report provided us with an updated call to action. The world's leading scientists have once again shown that we have no time to waste and that, as António Guterres, the Secretary-General of the United Nations, stated, we must now do "everything, everywhere, all at once" if we are to adhere to the Paris agreement's 1.5°C temperature limit. The next IPCC assessment cycle may not be concluded until 2030; this is the same timeframe in which we must make sufficient changes to prevent exceeding a critical threshold that, if crossed, will result in large and irreversible changes to the climate system.

Digital products and services underpin nearly all of our daily activities; they are everywhere, and the time has come to focus on reducing their energy consumption, if not all at once, as quickly and as broadly as possible.

After speaking with some of the world's leading cloud service providers, software companies, digital leaders, technology consultancies, and system integrators, I am convinced that we have an opportunity–one that could propel digital to the forefront of decarbonisation.

Our industry's place in the future

The technology industry is bursting with ingenuity, innovation, and problem-solvers. We must stop believing that taking action will require us to slow things down and return to a less polluting version of the past. In our digital world, we must move forward, maintain a progressive stance, and utilise technological advances to create the efficiencies required to support the continued growth of digital.

But we must recognise that we can only meet the growing technological needs of societies and businesses if we base this expansion on a more sustainable IT sector. As you will have seen in this book, we have everything we need to start making the changes needed to drive action. Please seize this opportunity to make a difference.

Creating capacity in the system

I have been in too many meetings where intelligent people have stated, "Yes, but once everything is running on renewable energy, everything will be fine, so we can continue doing what we are doing now, right?"

Let's reiterate why this reasoning is flawed.

If you've read the book all the way through, you'll understand why reduction is and will remain our top priority for the next decade. Renewables are critical, but it will take time for them to reach the required levels. Global renewable energy consumption is currently estimated at 29 percent[165] and is only projected to reach 35 percent by 2025[166]. To limit global warming to

[165] Source: Centre for Climate and Energy Solutions

[166] Source: International Energy Agency (IEA)

1.5°C, renewables must account for 61 percent of global electricity[167]. So there is a massive gap to bridge ahead of us. Alongside this, the impact of new technologies such as carbon capture remains controversial, with optimistic observers claiming that "carbon capture can achieve 14 percent of the global greenhouse gas emissions reductions needed by 2050"[168]. Again, while significant, the length of time needed suggests that it is unlikely enough of a contribution will be made to prevent the 1.5°C tipping point from being reached by 2030.

All of these new technologies will require even more energy for production, scaling, and implementation. Energy that will continue to rely heavily on fossil fuels. Even in a circular economy, we are merely extending the useful life of products to offset demand and the need to extract even more raw materials from the ground; we are not significantly altering or reducing consumption demand. So let's keep things straightforward. We must build capacity within the system to enable the testing, development, and widespread implementation of new technologies. To create this capacity, we must use what we already have in a significantly more efficient manner while also reducing our consumption wherever possible. The beautiful thing about digital is that we can begin decarbonising it using our current skills and capabilities; we just need to better inform ourselves about where to focus and then decide to take action.

The power of the reader

I have the good fortune to be invited to write thought leadership articles, speak at events, and occasionally serve as a panellist when they require someone with knowledge of decarbonising IT. I like to end my lectures with a message to the audience, which I refer to as the 'Power in the Room', and I

---

167 According to the World Resources Institute (WRI)

168 Source: Centre for Climate and Energy Solutions

think it is appropriate to attempt to give the same message to you, the reader:

"Thank you for reading this book. Please consider the following before leaving: I believe we have a generational opportunity–a chance as technologists to use our expertise to make a difference. Every reader of this book can contribute something. We can all see the need for change, and for those who design, build, operate, and manage digital products and services, we have everything we need to be a force for positive change beginning today. Use that power to make a difference and take action to reduce the carbon impact of digital."

Taking action involves choosing to do the right things and making the right decisions, interventions, and prioritisation calls at the right time in order to bring about positive change. Follow my recipe for change and start by measuring the carbon footprint of your technology and digital products and services as a way of mobilising your action. Use the Digital Decarbonisation Framework so that you can exert positive pressure on your own technology decisions and your ICT-related supply chain. By doing so, you will become part of a movement that drives the decarbonisation of digital.

In our digital world, we must be progressive in our actions and utilise technological advances to create the efficiencies required to support the growth of business and customer needs. But we must recognise that we can only meet the demand for more technology if this growth is based on a more sustainable IT sector.

To accomplish this, let's play our part in a global digital reset, and together we can create a lower-carbon digital future.

Afterword

I have been very privileged to be an observer, occasional advisor, and constant cheerleader for Eric and the team at GoCodeGreen, as they have gone from a start-up with a brilliant idea to a business recognised for its impact.

My interest in the consequences of technology on the planet started about the same time Eric was incorporating the company. I had a simple question. I was trying to answer an article I was asked to write for a publication. Is anyone focused on measuring the carbon footprint of technology? It turns out there was someone named Eric. We connected and chatted, and we have remained in close contact since.

As Eric has explained, we live in an age dominated by technology. Almost everything we touch and do involves hardware and software. We are surrounded by computers from the moment we wake up and look at our smart watches, to reaching for our phones to check social media, to the journeys we make in our cars, to our places of work, and to our entertainment. And the proliferation of machines will only increase as we move into a world where the Cloud and AI dominate.

Computing has a significant impact on the environment. On the scale we have become used to, it requires massive amounts of energy and water and has become a significant contributor to greenhouse gas emissions. We have to accept that digital technology is a fact of modern life, but if it continues to consume energy and water and generate excess heat in the way it is currently doing, its impact on our warming world will be devastating.

So, we have to find ways to understand the problem and make the technology footprint as small as possible.

Eric and the team at GoCodeGreen have codified a way of measuring the carbon footprint of software so that it is possible to make firm recommendations on how to make reductions. This has been such a positive and decisive move for the ICT industry. It shines a light on how correctly architecting, designing, and developing software can make a material difference to its carbon footprint.

This book provides so much insight into what that means and how, individually and collectively, the development community can adopt best practices and ensure that they work within an IT strategy that places the planet as a stakeholder.

If you work in IT, you should read this book. The ask is a gentle convincing currently, but regulation will come that mandates best practices for green development in the future. Now is the time to act and get ahead of the curve.

Honestly, that is what the planet needs. Less talk and more action!

Finally, Eric, as someone who has dyslexia, the manga is much appreciated!

Dave Wallace
Founder, Podcast Host, Writer and Advisor
October 2023

References

Page	Reference
3	Nasa Earth Observations, neo.gsfc.gov
10	European Parliament, 'The EU climate strategy for the ICT sector', Sept. 2022
10	Enerdata, 'Executive Briefs', 2018
13	Jonath Prac, manga artist, 2023
19	UK Parliament, 'Energy Consumption of ICT', Sept. 2022
19	GlobalData, 'Information Technology (IT) Sector ICT Market Size And Forecast (By Country, IT Solution Area And Vertical) To 2027', July 2023
19	Macrotrends, 'World Population Growth Rate 1950-2023'
19	DataCenter Knowledge, Apr. 2019
19	Dgtl Infra, 'Amazon Web Services (AWS) Data Center Locations: Regions and Availability Zones', June 2022
20	Statista, 'Global Figure of Hyperscale Data Centers 2015-202', Aug. 2023
20	International Energy Agency (IEA)
20	Oberlo, Q1 2023
20	Statista, 'Worldwide digital population 2023', Aug. 2023
20	Zippia, '75 Incredible email statistics', March 2023
20	Carbon Literacy Project, 'The carbon cost of an email: update!', Sept. 2022
21	Calculations based on data from www.pawprint.eco and www.climatelinks.org

21 According to research completed by Symantec in 2018

21 The Guardian, 'Bitcoin is terrible for the environment – can it ever go green?', April 2023

22 WhatsOn, '10 songs causing most CO2 emissions'

22 The Guardian, 'Streaming's dirty secret: how viewing Netflix top 10 creates vast quantity of CO2', 2021

22 Aleksa Gordic 'Leaked OpenAI data' from July 2023

25 Sources: UEA Earthshot prize nominations (uea.ac.uk) and Startups Magazine (www.startupsmagazine.co.uk), March 2023

27 climate.nasa.gov, 'Carbon Dioxide Vital Signs' Aug. 2023

27 MIT Climate Portal, 'How much carbon dioxide does the Earth actually absorb?', Jan. 2022

27 IFRF, '2011-2020 declared warmest decade on record', Jan. 2021

28 United Nations, un.org, 'Climate Action Fast Facts'

28 IPCC, '2021: Summary for Policymakers'

29 Source: World Health Organisation (WHO)

30 Climate Neutral Group, 'Five future scenarios' Climate Report AR6 IPCC, 2021

31 IPCC, Jim Skea, Working Group III Co-Chair, April 2022

32 BBC.com 'Smart guide to climate change', March 2022

32 SGR.co.uk using BEIS 2021 conversion factors

35 MeetGreen, meetgreen.com, 'Carbon Offsets - The Basics'

35 Financial Times, 'Cost of neutralising carbon emissions soars as demand escalates', per data from S&P Global Platts, Jan. 2022

36 McKinsey Sustainability, 'A blueprint for scaling voluntary carbon markets to meet the climate challenge', Jan. 2021

36 psmag.com 'How the green new deal can avoid climate colonialism', Feb. 2019

36 thehill.com 'Avoiding carbon colonialism: developing countries can't pay the price for pollution', April 2021

37 Fertile Ground - state of forest carbon finance 2017', Hamrick and Gallant, Dec. 2017

37 Oxfam case study: 'The new forests company and its Uganda plantations', Grainger and Geary, Sept. 2011

39 U.S General Services Administration, 'Embodied carbon: Next steps for GSA', July 2021, copyright Carbon Leadership Forum

40 therestartproject.org 'Mobiles: the global carbon footprint'

41 apple.com, 'Product Environmental Report: iPhone 14 Pro', Sept. 2022

41 Cult of Mac, referencing Omdia, Aug. 2023

41 Statista, 'Apple worldwide shipments of smartphones from 2010 to 2023, by quarter', Aug. 2023

41 energysavingstrust.org.uk, Oct. 2020

42 apple.com, 'Product Environmental Report: iPhone 15 Pro and Pro Max', Sept. 2023

43 Marine Georesources and Geotechnology, 'Environmental Impact Assessment of the Proposed Titanium Mining Project in Kwale District, Kenya', J.O.Z Abuodha, Nov. 2010

44 See: https://sciencebasedtargets.org/resources/files/Net-Zero-Standard.pdf

46 IPCC, Feb. 2022

46 public.wmo.int 'The Annual Global Carbon Budget', 2017

47 carbonindependent.org 'The global carbon budget will run out in 8 years', Sept. 2023

47 globalcarbonproject.org 'Global Carbon Budget 2021' infographic

50 Map sourced from Greenpeace, 'What is the Paris climate agreement and why does it matter', March 2021

51 United Nations Development Programme, 'The state of climate ambition', 2021

51 Organisation for Economic Cooperation and Development (OECD) (2021) and Oxfam International (2021).

53 Natural Resources Defense Council), nrdc.org, 'Greenhouse Effect 101', June 2023

54 carboncollective.co, 'Carbon Dioxide Equivalent', Sept. 2023

55 National Atmospheric Emissions Inventory, https://naei.beis.gov.uk/

56 GHG Protocol, ghgprotocol.org

57 WRI/WBCSD 'Corporate Value Chain (Scope 3) Accounting and Reporting Standard

60 Clim'Foot, 'Emission Factors', climfoot_project.eu

62 SDG: United Nations Development Programme website at undp.org

62 Science Based Targets Initiative: sciencebasedtargets.org

64 Components of ICT: TechTarget, techtarget.com

64 OECD, 'Measuring the Information Economy', 2002

65 Statista, 'Global ICT market share 2013-2023', July 2023

65 GeSI, 'ICT Sector Guidance Built on the GHG Protocol Product Lifecycle Accounting and Reporting Standard' - go to gesi.org for more information

66 Source: Dark Reading, R. Lemos, March 2022

70 ISG 'How Cloud Computing Reduces Carbon Emissions' - Not validated by the author.

70 TechTarget, 'A primer on hyperscale data centers', Feb. 2023

72 Source: Climatiq.io, CPU Emission factors via cloudcarbonfootprint.org and OEFDB

72 Climatiq, 'Clouding the issue: Are Amazon, Google and Microsoft really helping companies go green?', May 2022

72 GeoCurrents, Mapping Renewable Electricity Generation', Martin W. Lewis, using data from Wikipedia, geocurrents.info (undated)

75 Our World in Data, after Jones et al, 2023

76 UK Parliament, 'Energy Consumption of ICT', Sept. 2022

76 Assessing ICT global emissions footprint: Trends to 2040 & recommendations', Lotfi Belkhir and Ahmed Elmeligi, Jan 2018

77 UN SDG Goals - UN Statistics Division

82 Source: ourworldindata.org, referencing country level carbon emissions from 'Global Carbon Budget 2022'

83 The Register: 'Bill Gates on climate change: Planting trees is not the answer, emissions need to be zeroed out to avoid disaster', Feb 2021

94 Source: Network World, 'IT Best Practices', L. Musthaler, Dec. 2014

132 Source: httparchive.org 'Report: State of Images'

153 Find out more on Floyd's Cycle Detection Algorithm at geeksforgeeks.org

160 firebase.google.com, 'FCM Architectural Overview'

175 Source: hopl.info and wikipedia.org, 'List of Programming Languages'

176 CBLG: https://benchmarksgame-team.pages.debian.net/benchmarksgame/

189 EU Science Hub

192 Credits & Acknowledgements:
Throughout this section we will be referring to the Double Diamond Design Thinking approach, all credit and copyright is attributed to the Design Council per Attribution 4.0 International (CC BY 4.0).

203 Reference the Design Research Society, 'An Analysis of Design Maturity Models Used in Design Organisations' - Giri and Stolterman, June 2022

206 Antoine de Saint-Exupery, from 'Airman's Odyssey', 1942

207 almanac.httparchive.org 'Annual State of the Web Report', 2022

211 Articles such as 'How to Fight Climate Change with Open Source' at fosslife.org

212 Greenpeace, 'Clean Cloud 2021' report

212 Green Web Foundation and Enterprise Apps Today

219 Accessibility is Good for the Planet', see openweb.eu.org, June 2014

220 Reference: www.w3.org World Wide Web Consortium: 'Web Content Accessibility Guidelines'

223 Statista, 'Global number of digital video viewers 2019-2023', May 2023

224 Source: theshiftproject.org, 'Climate Crisis: the unsustainable use of online video', 2019

224 formafantasma.com, a collaboration with Studio Blanco

228 Helen Keller, 'My Story of my Life', 1902

238 Diagram from brightest.io, 'Understanding Scope 1, 2 3 Emissions'

240 Scope 3 emissions: Carbon Trust (carbontrust.com) and Carbon Disclosure Project (CDP) - cdp.net

244 GHG Product Life Cycle Accounting and Reporting Standard

251 SBTi Corporate Net Zero Standard, V1.1, April 2023

251 Simon-Kucher.com, 'Global Sustainability Study 2021', Oct. 2021

252 ey.com, '2021 EY Global Institutional Investor Survey' - 'Is your ESG data unlocking long-term value?', Nov. 2021

253 SBTi Corporate Net-Zero Standard (reference point 9 pertains to the inability of the maritime transport sector to set a target year for near-term goals prior to 2030).

255 Statista, 'Internet and Social Media Users in the World 2023', Aug. 2023

255 theshiftproject.org 'Environmental Impacts of Digital Technology: 5 Year Trends and 5G Governance'. Feb. 2023

261 Capgemini, 'Sustainable IT: Why it's time for a Green Revolution for your organisation's IT', 2021

263 Time, 'The Secret Cost of Google's Data Centres: Billions of Gallons of Water to Cool Servers', April 2020

263 MIT Technology Review, 2022

266 nationalgrideso.com and their 'Carbon Intensity Dashboard'

268 CGRi: The Circularity Gap Report 2023

270 World Economic Forum, 'A New Circular Vision for Electronics : Time for a Global Reboot', Jan. 2019

271 United Nations Environment Programme, 'Global Material Flows and Resource Productivity', 2016

271 CGRi: Circle Economy, 'The Circularity Gap Report 2021'

272 commission.europa.eu for details of the European Green Deal

273 tes-amm.com, 'Sustainability Report, 2021'

276 growyourowncloud.com

278 Science News, 'Why planting tons of trees isn't enough to solve climate change', Carolyn Gramling, Earth and Climate Writer, July 2021

278 The Conversation, Bonnie Waring, Grantham Institute - Climate Change and Environment, Imperial College, London, April 2023

278 The New York Times, Climate Forward Live Event, 'Discussing the Threats of a Rapidly Warming Planet', Sept. 2023

279 bonnchallenge.org, Forest Landscape Restoration

279 Annals of Forest Science, 'Implementing forest landscape restoration under the Bonn Challenge: a systematic approach', Stanturf, J.A., Kleine, M., Mansourian, S. *et al. May 2019*

281 Fatih Birol, IEA Executive Director, Nov. 2021

282 MIT Climate Portal, 'How much carbon dioxide does the Earth actually absorb?', Jan. 2022

283 Nasa Earth Observatory, 'Effects of Changing the Carbon Cycle', 2011

285 IUCN, 'Land degradation and climate change', Nov. 2015

286 National Geographic, 'Desertification, explained' C Nunez, May 2019

287 IUCN, 'The Ocean and climate change', Nov. 2017

287 IPCC, WGIIAR5-Chapter 30 'The Ocean', Hoegh-Guldberg, O., R. Cai, et al

288 United Nations, 'The Ocean - the world's greatest ally against climate change', (undated)

288 IUCN, 'Species and climate changes', Oct. 2021

289 IUCN 'Red List of Threatened Species™', see https://www.iucnredlist.org

289 BBC, 'Nature crisis: One in six species at risk of extinction in Great Britain', V. Gill and K. Stephens, Sept. 2023

290 IUCN, 'Forests and climate change', Feb. 2021

291 Go to https://www.wildlifetrusts.org/ for more information

292 UN Environment Programme, 'Earth's biodiversity depends on the world's forests', May 2020

292 nhm.ac.uk, 'How are climate change and biodiversity loss linked?', Chetan-Walsh and Hendry, (undated)

293 European Commission, 'Nature's role in climate change', Aug. 2009

294 Horizon, EU Research and Innovation, 'AI can help us fight climate change. But it has an energy problem, too', Sept. 2019

295 PWC report, 'How AI can enable a sustainable future' (undated)

295 Nature Communications, 'The role of artificial intelligence in achieving the Sustainable Development Goals', Jan. 2020

298 United Nations Environment Program, Digital Transformation sub-programme, 'How artificial intelligence is helping tackle environmental challenges', Nov. 2022

300 capgemini.com, 'Sustainable IT is the backbone of a greener future' and 'Sustainable IT: Why it's time for a Green Revolution for your organization's IT' (undated)

301 Maria Mendiluce, CEO, We Mean Business Coalition, 'We can't afford to let perfection be the enemy of companies taking climate action', Feb. 2023

302 World Resources Institute, IPCC AR6

304 The Conversation, 'Climate scientists: concept of net zero is a dangerous trap', James Dyke (University of Exeter), Robert Watson (University of East Anglia) and Wolfgang Knorr (Lund University), April 2021

304 IEA, 'Energy Efficiency 2021', Nov. 2021

305 IEA, 'Energy Efficiency 2018, Analysis and outlooks to 2040', Oct. 2018

308 James Heath, CEO National Infrastructure Commission, July 2023

321 Organizational World', Howard J Leavitt, 1973

324 The 'DevSustainOps: The management practice of embedding and integrating sustainability into the delivery, development and operation of digital products and services', Version 2.0, March 2023 is available by contacting GoCodeGreen

326 World Business Council for Sustainable Development (WBCSD)

326 Ecovadis. www.ecovadis.com

330 IT Sustainability Think Tank: Keeping on top of the regulatory landscape', Richard Barnes-Webb (PA Consulting) and Eric Zie (GoCodeGreen), Computer Weekly, July 2023

325 IFRS S1 'General Requirements for Disclosure of Sustainability-related Financial Information', June 2023

351 IFRS S1 (International Financial Reporting Standards): General Requirements

352 IFRS S2: Climate-related Disclosures

352 ESRS E1

355 Deloitte Insights, 'Cut carbon, cut costs', Feb. 2022

358 britannica.com, 'Conditional Probability - mathematics', Stephen Eldridge, Sept. 2023

369 Harvard Business School, 'Sustainable Business Strategy', HBS Professor Rebecca Henderson, (undated)

370 McKinsey Sustainability, 'Organising for sustainability success: Where, and how, leaders can start', Aug. 2021

380 The Lancet - Planetary Health, 'Climate anxiety in children and young people and their beliefs about government responses to climate change: a global survey', Dec. 2021

380 TheClimateOptimist.com, Anne Therese Gennari, Founder

381 IBM Institute for Business Value, 'Balancing Sustainability and Profitability: Consumer research highlights how businesses can protect people, planet and the bottom line', April 2022

382 Reuters, April 2022

383 capgemini.com, 'Reinventing work and sustainability' (undated)

393 Source: Centre for Climate and Energy Solutions

393 Source: International Energy Agency (IEA)

394 According to the World Resources Institute (WRI)

394 Source: Centre for Climate and Energy Solutions

Index

This is an illustration of the index. It is my visual checkpoint at the end of the book to ensure that the green (action-oriented, positive topics) outweighs the red and amber challenges related to climate change. This is the book I wanted, one of positive optimism.

D

E

S

U

V

W

Y

Stop Press

The first proof copy of this book was sent to the printer on the evening of October 6, 2023. The following morning, I awoke to the following BBC headline:

"World breaches key 1.5C warming mark for record number of days"

Each passing day is bringing us closer to irreversible climate change. Let's use the information in this book to start taking action.

Stop Press (again)

Just in time for the second proof edition, Storm Babet hit the UK on Friday, October 20, 2023. My street was flooded, and Belle was not amused. This is yet another sign that climate change has literally hit our doorsteps.

I started my journey to decarbonise digital nearly three years ago, and people weren't yet ready to take on the challenge. I truly hope that today, as I stand on a street that has never been flooded before, there will be more of us able to see the opportunity to make a difference.

Literally every week, we hear more about our continued impact on the planet. We need a global digital reset. This is our challenge. Our opportunity.